California Serpentines:

Flora, Vegetation, Geology, Soils,

and Management Problems

by Arthur R. Kruckeberg

UNIVERSITY OF CALIFORNIA PRESS
Berkeley   Los Angeles   London

UNIVERSITY OF CALIFORNIA PUBLICATIONS IN BOTANY

Volume 78

Issue Date: December 1984

UNIVERSITY OF CALIFORNIA PRESS
BERKELEY AND LOS ANGELES,
CALIFORNIA

UNIVERSITY OF CALIFORNIA PRESS, LTD.
LONDON, ENGLAND

ISBN: 0-520-09701-7
LIBRARY OF CONGRESS CATALOG CARD NUMBER: 83-18237

Library of Congress Cataloging-in-Publication Data

Kruckeberg, Arthur R.
  California serpentines.

  (University of California publications in botany;
v. 78)
  Bibliography: p.
  Includes index.
  1. Botany—California—Ecology. 2. Serpentine—
California. 3. Rocks, Ultrabasic—California. I. Title.
II. Series.
QK149.K78   1985   582'.05   83-18237
ISBN 0-520-09701-7 (alk. paper)

Printed in the United States of America

08 07 06 05 04 03 02 01
      9 8 7 6 5 4 3

The paper used in this publication meets the minimum requirements of
ANSI/NISO Z39.48-1992 (R 1997) (*Permanence of Paper*). ⊚

CALIFORNIA SERPENTINES: FLORA, VEGETATION, GEOLOGY, SOILS AND MANAGEMENT PROBLEMS

Two classic serpentine localities in the Coast Ranges of California.

Above: New Idria area, upper Clear Creek, San Benito County. (Photo by J. Griffin.)

Below: Hill 1030 (USGS benchmark), ca. 4 miles northeast of Middletown, Lake County, along State Highway 29. (Photo by author.)

To my father, Arthur Woodbury Kruckeberg, whose enthusiasm for California plants infected me for life, and to Herbert L. Mason, who sowed in me the germ of fascination for the plant life of serpentines.

CONTENTS

LIST OF ILLUSTRATIONS

LIST OF TABLES

ACKNOWLEDGMENTS

In the beginning there was the inspiration of Herbert L. Mason, who knew the significant role of edaphic discontinuities in the distribution and diversity of the California flora. Then further encouragement came from G. Ledyard Stebbins, Jr., and Hans Jenny. These three Berkeley scientists started me along the serpentine trail. Other Bay Area botanists gave of their special talents over the years: Rimo Bacigalupi, Annetta Carter, Lincoln Constance, Larry Heckard, Alice Howard, John Thomas Howell, and Robert Ornduff.

Freed Hoffman must be especially memorialized. An amateur botanist, he knew the most about the plant life of California serpentines and could theorize heuristically on causes and relationships, free from academic restraint.

I am especially grateful to my colleague, Richard B. Walker, whom I followed to Seattle from Berkeley in 1950; he has always provided wise and cautious counsel about solutions to the physiological enigmas of serpentine addiction.

Pre-publication costs were defrayed in part by generous grants from the California Native Plant Society and the University of Washington Graduate Research Fund, for which I am most grateful.

Grant support has come over the years from the University of California Graduate School, the University of Washington Initiative 171 and Graduate Research funds, and from three National Science Foundation Grants (G-1323, GB-4579, and GB-11710). In between grants, I apportioned funds to my serpentine diversions out of annual teaching income. This academic who gets paid for doing what he loves to do is ever grateful to the University of Washington for sustained support over the years.

ABSTRACT

The diversity of California's natural environments includes over 1100 sq. mi. of ultramafic outcrops (serpentinite and other ferromagnesian rocks). Endemism, indicator species, and sparse vegetation cover reflect the severity of the serpentine habitat.

Other than William Brewer, early California botanists failed to record impressions of the association of unusual plant life with serpentine soils. Mason in the 1940s was the first to recognize the close correspondence of plants to serpentines. The floras of the state (Abrams, Munz and Keck) often record the preferences of certain plants to serpentine; Jepson's two major works on the flora do not refer to the serpentine habitat. Soil scientists and plant physiologists attempted to relate the chemical nature of serpentine soils to plant growth in the 1920-40 period.

California geologists recorded the presence of serpentines beginning early in the 19th century. Economic geologists of the early 20th century recognized the indicator value of serpentine outcrops and their barrenness in searching for quicksilver and other minerals. Serpentine outcrops are largely distributed in the foothills of the Sierra Nevada and in the Coast Ranges from San Luis Obispo County north to the Oregon border. Ultramafics are frequently associated with the Jurassic Franciscan formation of California.

Ultramafics consists of iron-magnesian silicates, with olivine, chromite, and pyroxene as the major minerals. Serpentinites contain the minerals chrysotile, lizardite, and antigorite; iron and magnesium are also chief elements of these minerals. Ultramafic outcrops may have been available for plant colonization at least through the late Tertiary (Miocene to Pleistocene). Serpentine and other ultramafics weather to soils rich in magnesium, iron, and silicates. Nickel often occurs in exceptional amounts. Smectite (montmorillonite) is the typical clay mineral of serpentine soils. California soil series derived from serpentinite include the Henneke, Dubakella, and Montara series from upland soil types; four alluvial soils of serpentine origin have been named. The upland soils are shallow and stony, with little profile development (skeletal or azonal soils).

Serpentine soils have high values of exchangeable magnesium and exceptionally low values of exchangeable calcium. They are usually deficient in nitrogen and phosphorus; molybdenum deficiency also can occur. The heavy metals, especially iron and nickel, can attain high values in both soils and plant tissues. Hyperaccumulation (>1000 ppm) of nickel by a serpentine endemic, *Streptanthus polygaloides*, has been reported. The causes of the "serpentine syndrome" have been traced to imbalance of calcium and magnesium, magnesium toxicity, heavy metal toxcity, or low levels of essential nutrients. A more likely explanation involves several factors: the interaction of stresses in chemical composition of soils, the physical features of the habitat, and the biological consequences of inadequate nutrient recycling.

Vegetation of California serpentines contrasts sharply in physiognomy and species composition with vegetation on adjacent non-serpentine soils. The shift in vegetation type and life-form across a serpentine contact often involves the replacement of oak woodland with serpentine chaparral, Douglas fir and hardwoods with pine-cypress, or chaparral with sparse grassland. Extreme serpentine sites may be nearly devoid of plant life. Serpentine chaparral consists of endemic shrubs (*Quercus durata, Ceanothus jepsonii,* and *Garrya congdonii*) as well as non-endemic chaparral species. On serpentines in northwestern California, open woodlands (of Jeffrey pine and incense cedar) replace the typical mixed conifer-hardwood forests of yellow pine, Douglas fir, redwood, etc.

Taxonomic and evolutionary responses of the California flora to serpentines include (1) taxa endemic to serpentine; (2) local or regional indicator taxa, largely confined to serpentine in parts of their ranges; (3) indifferent or "bodenvag" taxa that range on and off serpentine; (4) taxa that are excluded from serpentine.

Two hundred and fifteen species, subspecies, and varieties of the native flora are restricted wholly or in large part to serpentine soils; they include conifers, woody and herbaceous dicots, monocots, and ferns. Examples of endemics: trees–*Cupressus sargentii* and *C. macnabiana*; shrubs–*Quercus durata, Ceanothus jepsonii,* and *Arctostaphylos franciscana*; herbaceous perennials–*Calochortus tiburonensis, Fritillaria falcata, Arabis macdonaldiana, Senecio greenei, Monardella benitensis,* and *Galium hardhumiae*; many annuals–*Streptanthus* (13 species of subgenus Euclisia), 8 species of *Hesperolinon, Layia discoidea, Benitoa occidentalis,* and *Lessingia ramulosa.*

Serpentine indicators are either faithful to serpentine only in parts of their total range (e.g., *Pinus jeffreyi, Calocedrus decurrens,* and *Darlingtonia californica*) or are indicators on serpentine throughout most of their total ranges as recruits from local xeric non-serpentine habitats (*Pinus sabiniana, P. attenuata, Ceanothus cuneatus, Adenostoma fasciculatum,* etc.).

Indifferent or bodenvag taxa are found on both serpentine and adjacent non-serpentine habitats. Some are ecotypic, with serpentine-tolerant and intolerant races (e.g., *Achillea lanulosa, Gilia capitata, Salvia columbariae,* etc.).

Besides a number of woody native species that avoid serpentine (*Sequoia sempervirens, Abies concolor, Quercus agrifolia, Cercocarpus betuloides,* etc.), herbs of many different genera are absent on serpentine (e.g., *Penstemon, Ranunculus, Aster, Erigeron, Lotus, Lupinus, Oenothera, Trifolium,* etc.).

Little is known about faunal responses to serpentines. Specialized herbivory of serpentine plants is reported for butterflies (*Euphydryas editha* and *Pieris* spp.). Serpentine species of the crucifer, *Streptanthus,* respond to picrid predation by a kind of egg mimicry.

Since adaptation of plants to serpentines ranges from racial tolerance to monotypic endemism, the origins of such adaptation are likely to have been varied. Explanations of serpentine adaptation include (1) ecotypic differentiation, (2) "drift" and/or depletion of biotypes, (3) catastrophic selection and saltational speciation, (4) gradual allopatric speciation with ecogeographic specialization, (5) hybridization with or without polyploidy. The taxa of subgenus Euclisia (*Streptanthus*–Cruciferae) show restriction to serpentine in varying degrees, from local tolerant races and subspecies to wide-ranging serpentine endemics and very narrow, local endemics.

Modification of serpentine habitats by human activity results from mining activities, agriculture (farming and grazing), logging, and recreation. Some disturbances (mining, logging, and the development of geothermal power) have affected serpentine floras. The least destructive uses of serpentine areas are as watershed and by wildlife; the infertility of serpentines limits their value for agriculture and forestry.

Preservation of serpentine biota, so rich in endemics, is needed in California. Very few natural areas have been set aside as serpentine preserves. Incidental preservation occurs when state and federal lands that have serpentines are set aside as parks, etc. (e.g., Mt. Diablo, Mt. Tamalpais). Only one serpentine preserve is under U.S. Forest Service jurisdiction; other state and federal holdings receive protection by "benign neglect."

A number of serpentine endemic taxa are rare enough to be on the U.S. and California lists of rare and endangered species. Preservation of such target species also affords protection for the serpentine habitats where they occur.

INTRODUCTION

The rich and highly endemic flora of California epitomizes the plant geographer's dictum: Floristic diversity is fashioned out of environmental diversity (Stebbins and Major, 1965; Raven and Axelrod, 1978). A major component of California's ecological diversity stems from the workings of a great array of geologic processes and products. Different landforms resulting from a variety of geomorphic processes, and many different rock types of sedimentary, metamorphic, and igneous origin, plus the consequent soil-forming processes, have created a lavish diversity of habitats within the varied climatic regimes of the state.

Though Cain (1944) has insisted that edaphic control of plant distribution is secondary to climatic controls, the effect of soils and kindred geologic factors must be accorded great importance in the origin and diversity of the California flora. For here geologic processes and materials are far from monotonously uniform. "California is a state of geologic contrasts. Of the 48 contiguous states, it contains the highest and lowest elevations only 80 miles (130 km) apart, plus a variety of rocks, structures, mineral resources, and scenery equalled by few areas of the world. . . . California's rocks vary from ancient Precambrian to presently forming sediments, and several of the state's formations are type examples for North America and the world" (Norris and Webb, 1976, p. 1).

Common ingredients within this rich geologic tapestry are ultramafic[1] rocks, largely serpentinite and peridotite, which weather to soils of exceptional physical and chemical properties. Most of the peridotites are completely or partially serpentinized, and therefore serpentinite is the most ubiquitous rock of the ultramafic clan in California, covering 1100 square miles throughout much of the state. Both peridotite and serpentinite profoundly alter the pattern of vegetation and species composition of the vascular flora nearly everywhere they occur (Krause, 1958; Kruckeberg, 1954, 1969a,b; Ornduff, 1974; Proctor and Woodell, 1975; Rune, 1953; Whittaker, 1954). A most telling effect is on the vegetation, particularly its physiognomy. Oak savannah or mixed conifer forest on normal soils abruptly gives way to xerophytic scrub (= serpentine chaparral: Hanes in Barbour and Major, 1977) on serpentine soils. Notable in the xerophytic scrub on serpentinites[2] are endemic and indicator species like *Quercus durata* Jeps., *Garrya congdonii* Eastw., *Cupressus sargentii* Jeps., *Ceanothus jepsonii* Greene, and *Arctostaphylos viscida* Parry, in addition to more generally occurring chaparral species that are tolerant of serpentine.

Serpentines in California are noted for harboring endemic species of all life-forms. Some are narrowly restricted to a single or a few neighboring serpentine outcrops: *Streptanthus niger* Greene and *Calochortus tiburonensis* Hill on the Tiburon peninsula (Marin County);

1 *Ultramafic*: Ferromagnesian rocks containing more than 70% mafic [Mg Fe] minerals (Wyllie, 1967).

2 Hereafter the general term *serpentine*, including unserpentinized peridotite, will be used for rock, soil, and vegetation of ultramafic affinity.

Streptanthus batrachopus Morrison on Mt. Tamalpais; *Streptanthus brachiatus* Hoffman in the Big Geysers country of Sonoma and Lake counties. Other species endemic to serpentine have more extensive ranges: e.g., *Cupressus sargentii* from Santa Barbara County to Tehama County; *Ceanothus jepsonii* from Santa Clara County to Lake County; *Streptanthus breweri* Gray from Stanislaus County to Colusa County. The contrasts between vegetation on normal soils with that on serpentines are bold and sharp; even the most casual observer is drawn to the conclusion that the substrate has a profound effect on the plant life that becomes established on the serpentine "barren" and yet is absent on nearby nonserpentine sites.

I will review here what is known about the plant life of serpentine and other ultramafic substrates in California. After brief accounts of the history of botanical observation on serpentine, and the geology of California ultramafics, I will discuss the particular effects of serpentine and related soils on vegetation (mineral nutrition, morphological and physiological adaptations, etc.), community types on serpentine, and will provide an inventory of plants that grow on serpentine. Since serpentine floras are unusually rich in endemic taxa, consideration of their evolutionary origins forms a significant chapter. Finally, California serpentines have a history of exploitation— for resource extraction, recreation, etc.—and we need to examine the problems of land management associated with serpentines in California.

HISTORY OF BOTANICAL OBSERVATIONS ON THE SERPENTINE FLORA OF CALIFORNIA

In a fruitless search through journals of early explorers, I encountered no mention of a plant restricted to serpentine soil until the published records of William H. Brewer's botanical explorations of California (1861). Of that narrow serpentine endemic, *Streptanthus polygaloides* Gray, Brewer states (in A. Gray et al., 1880)· "A rare and remarkable species . . . on dry barren magnesian soil near Jacksonville on the Tuolomne." In Brewer's (1949) delightfully written recollections of this trek, *Up and Down California*, he records his impressions of that most striking of serpentine barrens, the country around the New Idria quicksilver mines: "The view from the summit is extensive and peculiar . . . chain after chain of mountains, most barren and desolate. No words can describe one chain, at the foot of which we had passed on our way—gray and dry rocks or soil, furrowed by ancient streams into innumerable canyons, now perfectly dry, without a tree, scarcely a shrub or other vegetation— *none*, absolutely, could be seen. It was a scene of unmixed desolation, more terrible for a stranger to be lost in than even the snows and glaciers of the alps" (pp. 139-140).

All during the period from 1880 to 1940, when the luminaries of California botany were busy describing new species, there was a singular lack of attention to the edaphic aspects of plant habitat. I find nothing in the writings of E.L. Greene, K. Brandegee, M.E. Jones, A.A. Heller, W.L. Jepson, or A. Eastwood to indicate that they had observed the striking contrasts between the serpentine barren and the adjacent normal substrates with more substantial plant cover. E.L. Greene (1904) devotes several pages to a review of cruciferous "genera" that he carved out of *Streptanthus*, without mention of the singular addiction to serpentine of at least two of his "genera." Neither in the Manual (1925), nor in the more scholarly Flora (1909 to 1943), does W.L. Jepson identify any taxa with special affinities for unusual substrates— serpentine or any other. Even the two most characteristic shrubs of California serpentine, *Quercus durata* and *Ceanothus jepsonii*, are simply described as occurring on dry hills and flats. Since it was Jepson who first recognized and published *Q. durata*, it seems strange that he failed to call attention to its serpentine specificity, but I searched in vain through the pages in the Manual where he presents an otherwise valuable introductory section on plant distribution and endemism. It is true that he recognized the importance of edaphic variation (1925, p. 18). But there is no mention of examples of the most striking edaphic types—serpentine endemics or the plants of acid sands. Professor Jepson's unpublished field notes tell of his wonderment at viewing (in May 1907) the wasteland of New Idria: "Great areas on the summits and ridge slopes are as barren as one's hand, not even herbaceous vegetation. These characteristic spots are rotting sliding shale rock . . . ." (in Griffin, 1975, p. 7). Jepson's mistaking serpentine for shale is perhaps understandable: both rocks fracture and cleave readily to make a flaky, loose talus.

Abrams' regional flora (1923–1960) seems to have been the first to specify a serpentine substrate for certain species. Abrams began the practice in volume 1 and continued to do so in all subsequent volumes, as did Roxanna Ferris in the volume on Compositae. Even so, the Abrams work fails to mention substrate for some of the obvious serpentine endemics like *Quercus durata* and *Ceanothus jepsonii*. In their preoccupation with describing new species and the writing of floras, the early California botanists apparently had no predilection for recording habitat preferences; an ecological view of the state's flora had yet to be developed.

If early California taxonomists overlooked the bold manifestations of serpentine vegetation, surely the pioneer phytogeographers might have been sufficiently impressed to have commented on the phenomenon in their writings. Harshberger (1911), in his phytogeographic survey of the whole North American continent, surely was on serpentine in northern California, for he wrote about the cypress endemic to serpentine as follows: "The eastern and western sides of Red Mountain [Mendocino or Glenn Counties?] are peculiar in the presence of groves of *Cupressus macnabiana*. . . . At Red Mt. grow *Lilium pardalinum, Aquilegia truncata, Epipactis gigantea*, and a fern, *Pellaea densa*" [= *Aspidotis densa*, a faithful indicator of serpentine]. Yet he never attributes the peculiar occurrence of the cypress to any cause, substrate or otherwise.

Cooper's early landmark paper (1922) on California chaparral just misses being the first major essay on serpentine plant life of California. Cooper chose to concentrate his study of the ecology of chaparral vegetation at Jasper Ridge, just west of Palo Alto. Though much of Jasper Ridge is serpentine, Cooper selected his plots on adjacent sandstone.

The soil science literature of the 1920s shows stirrings of interest in the serpentine problem. In 1926 Professors Gordon and Lipman of the University of California (Berkeley) published a study dealing with the infertility of California serpentine soils, a subject to which we will return later. United States Department of Agriculture soil scientists were concerned with the problems of infertility on serpentine too. Robinson and his colleagues published their chemical analyses of serpentine in 1935; among the samples tested were soils from Mt. Tamalpais and southern Oregon.

During the 1940s there was a flurry of interest in California serpentines. John Morrison's doctoral thesis (1941) on the taxonomy of the Euclisia section of *Streptanthus* emphasizes the strong affinity for serpentine that most of its taxa have for serpentine in central and northern California. Helen Sharsmith, another Berkeley botanist, stresses in her floristic study of the Mount Hamilton Range (1945) the importance of serpentine to endemic elements of the Coast Range flora. In that same decade, Herbert Mason brought the problem of narrow endemism and serpentine outcrops into sharp focus. His two papers (1946a,b) spurred a postwar generation of ecologists, physiologists, and soil scientists to work on aspects of the serpentine problem. Out of the Mason stimulus came my studies (Kruckeberg, 1951, 1954) and those of McMillan (1956), Vlamis and Jenny (1948), Vlamis (1949), Walker (1948, 1954), and Whittaker (1954, 1960).

Since the 1950s, only a few research papers on serpentine floristics and vegetation of California have appeared: Forde and Faris, 1962; J.T. Gray, 1979; Griffin, 1965, 1974,

1975; Hardham, 1962; Kruckeberg, 1957, 1958; McNaughton, 1968; Vogl, 1973; Waring and Major, 1964; Wells, 1962; B. Zobel, 1952; and D.G. Zobel and Hawk, 1980. Reviews of "the serpentine syndrome" (Jenny, 1980) include the Whittaker, Walker, and Kruckeberg symposium of 1954; Krause, 1958; Kruckeberg, 1969a; Proctor and Woodell, 1975; Barbour and Major, 1977; Raven and Axelrod, 1978; and Jenny, 1980.

GEOLOGY OF SERPENTINE AND RELATED ULTRAMAFIC ROCKS

Serpentinite is California's state rock!

LITHOLOGY AND MINERALOGY

Ultramafic rocks like serpentinite, dunite, and peridotite have in common some form of ferromagnesian silicate mineral. Such rocks occur on every major landmass of our planet, in local outcrops or as extensive regional displays. In western North America, the three Pacific coast states and the Canadian province of British Columbia have both local and massive occurrences of ultramafics—California with 1100 sq. mi. (2860 km²), Oregon with 450 sq. mi.(1170 km²), and Washington with 200 sq. mi. (520 km²). The chief constituent, some mineralogical variant of iron–magnesium silicate, gives ultramafic rocks their chemical and physical character and is very likely the major cause of the consequent biological properties of the outcrop.

In the Coast Ranges the ultramafics are largely serpentinite (see fig. 1), derived from ophiolites[1] of Mesozoic Age and companion to the vast and complex Franciscan sedimentary rocks. In the Sierran foothills, serpentinite is usually associated with sedimentary and volcanic rocks (ophiolites) older than the granitics of the Nevadan orogeny—the great series of intrusions that made the granitic Sierra Nevada. Though much disputed earlier in this century, there is now substantial agreement as to how the serpentinite invaded the sedimentary rocks. For Franciscan-associated serpentinites, Norris and Webb (1976, p. 253) state:

> All these Franciscan rocks have been intruded by ultrabasic igneous rocks, now serpentinized peridotite (serpentinite). Sometimes the serpentinites have been injected as normal molten intrusives, but in other instances they occur in sill-like sheets that lack the thermal alteration of enclosing rocks expected in most sills. In still other cases, these plastic serpentinites have squeezed up through the overlying rocks as plugs or diapirs [injections by a mobile core of a brittle overlying rock]. The prevailing view is that these serpentinized peridotites are altered masses derived from the upper mantle and transferred tectonically to the earth's surface.

R.G. Coleman, an authority on ultramafic geology, takes exception (personal communication) to portions of Norris and Webb's explanation. Rather than intrusion, Coleman uses the term "tectonic emplacement" to account for the appearance of ultramafics in Franciscan rocks. He also modifies Norris and Webb's last sentence above on mode of emplacement, to:

---

1 *Ophiolite*: An assemblage of mafic to ultramafic rocks of the oceanic crust and upper mantle (Coleman, 1977).

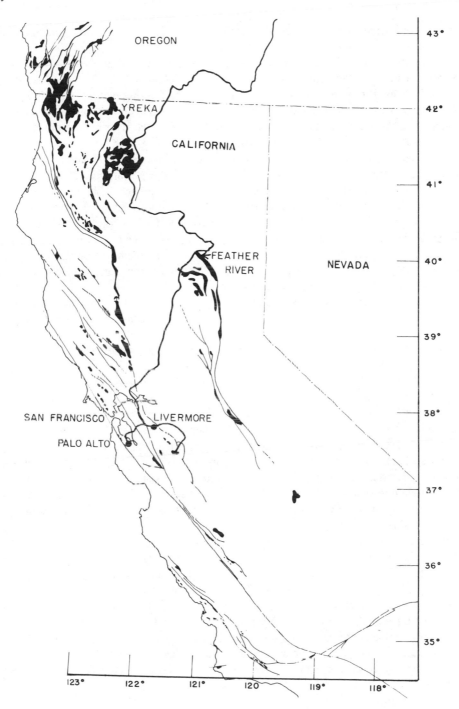

FIG. 1. Map of serpentine occurrences in California and Oregon. 1 cm = 55 km. (Courtesy of R. Coleman.)

" . . . serpentinized peridotites are altered ophiolites derived from the oceanic lithosphere and transferred tectonically to the continental margin."

Serpentinite and other ultramafic rocks tend to be associated with extensive southeast-to-northwest-trending fault zones (fig. 1) in both the cismontane Sierra Nevada foothills and the Coast Ranges. It now seems quite certain that both the faulting patterns and ultramafic intrusion are structurally associated with crustal movements (plate tectonics), as the ultramafic mantle of the oceanic crust moves under the continental crust. Ultramafics thus emerge at the margins of oceanic and continental plates.

Less common than the metamorphic rock serpentinite are the outcrops and intrusions that remained unaltered (igneous). Dunite, hartzburgite, and especially peridotite, three common igneous ultramafic rocks, are more abundant in the Klamath–Siskiyou region— the Mt. Eddy massif (Siskiyou County) and the adjacent mountainous terrain are largely peridotite.

The main mineral constituent of igneous ultramafic rocks is olivine: $(Mg,Fe)_2SiO_4$. But several other mineral types, such as chromite and pyroxene can occur in igneous ultramafics (Wyllie, 1967). The ultramafic rock serpentinite contains members of the serpentine-group minerals (Faust and Fahey, 1962) that are formed anew during metamorphosis: chrysotile (or asbestos), lizardite, and antigorite are the most prevalent of these minerals. No new elements absent in the parent igneous ultramafics are added; however, there may be some gain or loss of silica, magnesium, calcium, iron, or carbon dioxide. At the contact zones between igneous and metamorphic ultramafics, reaction-zone mineralization can take place. Coleman (1967) refers to the occurrence of rodingite at such contact zones; though recorded here in California, the phenomenon may be worldwide in occurrence (rodingite derives its name from its type locality in New Zealand). That rodingite is rich in calcium silicate minerals is significant in tracking the path of serpentine-tolerant plants across such a contact.

It is well to emphasize the rather wide range of variation in kind and amount of elemental and mineralogical constituents of ultramafic rocks, as well as the frequent occurrence of "contaminants" along contact zones. The consequences of this variation become apparent when ultramafics weather; the soils will invariably contain more than iron and magnesium silicates. Besides iron, magnesium, and silica, both soils and rocks of the ultramafic clan can contain elements such as calcium, aluminum, nickel, chromium, and magnesium, either in trace amounts or at more substantial concentrations (Faust and Fahey, 1962; Robinson, Edgington, and Byers, 1935).

Of great importance to the botanist is the geological timespan during which ultramafic deposits have been at the surface, available to plant colonization. Authors on California serpentinites differ on this matter. Coleman (1967, p. 5) states that "The presence of detrital serpentine debris within Mesozoic and Tertiary sedimentary rocks of California attests to the long exposure of ultramafic rocks at the surface." This implies their continued availability to plant life through the Tertiary. But Raven and Axelrod (1978) give a more recent time frame: they suggest that continuous exposure began in the Miocene for some localities, and for other areas in California came as late as the Pleistocene. Naturally such different time periods for exposure have to be considered when postulating the biological event of colonization on serpentine.

DISTRIBUTION OF ULTRAMAFIC OUTCROPS IN CALIFORNIA

Serpentinites and other ultramafic rocks occur in many places in California, from Santa

Barbara County to the Oregon border in the Coast Ranges and intermittently along the western foothills of the Sierra Nevada from Tulare to Plumas counties (fig. 2). I know of only one isolated outcrop south of Santa Barbara County, in the Santa Ana Mountains of Orange County (Vogl, 1973).

The following summary of serpentinite distribution in California (and southwestern Oregon) is taken from separate area sheets of the Geologic Map of California (California Div. of Mines and Geology, 1958–1967). The Fresno and Santa Rosa sheets illustrate well the elongate, south-to-north pattern of the outcrops noted above. (A more detailed account of the geography of serpentinite localities is given in Appendix A.)

South Coast Ranges

Going from south to north, the first substantial serpentine outcrops are encountered in the Figueroa Mountain area of the San Rafael Mountains in Santa Barbara County. The next areas of significance are in San Luis Obispo County, near the coast and inland at the eastern end of the Carrizo Plain. From here northward in the south Coast Ranges to the San Francisco Bay Region, serpentinite outcrops with great frequency. The most impressive of these outcrops are in the inner south Coast Ranges of San Benito County, notably in the New Idria barrens and San Benito Peak (1597 m altitude). (See figs. 3 and 4 for historic views of the area.)

San Francisco Bay Region

Serpentines are rather common in the Bay Region. They include the Crystal Springs area of San Mateo County, the Presidio within the city of San Francisco, and areas in the Oakland–Berkeley hills. Many of these outcrops are obscured by human development.

North Bay Counties (Marin, Sonoma, and Napa counties)

Here serpentines begin at sea level (Angel Island, Tiburon Peninsula) and can reach nearly to the summit of such peaks as Mt. Tamalpais and Mt. St. Helena. Although both local and extensive ultramafics are found in the southern sector of the region, the most spectacular and botanically rich outcrops are farther north, in northern Sonoma and Napa counties (figs. 5 and 6).

North Coast Ranges (Lake and Mendocino counties north to Josephine and Curry counties of southwestern Oregon)

From the arid foothills bordering the western edge of the Sacramento Valley to the fog belt of the coastal redwood region, serpentines outcrop with great frequency (figs. 7 and 8). The Klamath–Siskiyou mountains to the north have the largest ultramafic outcrops in the state, indeed in North America. From arid to mesic, from low elevation (Gasquet, Del Norte County) to the summits of the highest peaks (Mt. Eddy, 9038 ft. (2755 m), Dubakella Mountain, 5891 ft. (1796 m), and Preston Peak, 7310 ft. (2229 m), there is an ultramafic outcrop (either igneous or metamorphic) for nearly every kind of terrain and exposure in northwestern California.

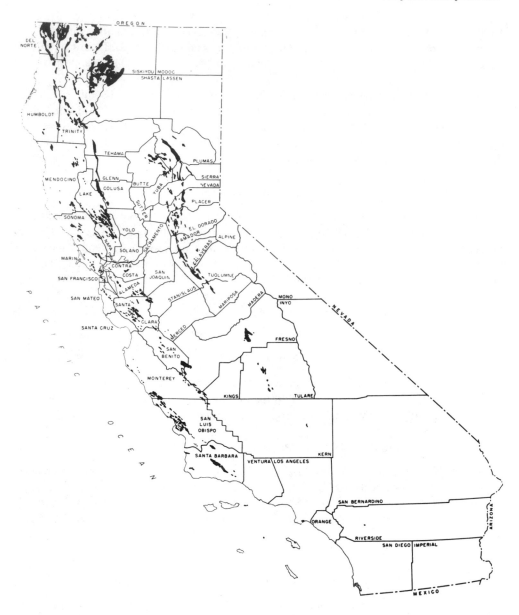

FIG. 2. Distribution of serpentine outcrops in California by county. 1 cm = 71 km. (With permission, California Division of Mines and Geology.)

FIGS. 3 and 4. Views of the stark, barren serpentine outcrops of the New Idria area, San Benito County. (Photos by A.E. Wieslander on May 7, 1932, courtesy of Pacific Southwest Forest and Range Experiment Station, U.S. Forest Service). Comparison with the view of the same area taken many years later (see upper photo of frontispiece), reveals little change in the barren scene.

FIG. 5. Massive serpentine outcrop in the vicinity of the Layton Mine, upper east Austin Creek, Sonoma County.

FIG. 6. Pure stand of *Cupressus sargentii* on ultramafics, locally known as "The Cedars," in the upper east fork of Austin Creek area, Sonoma County.

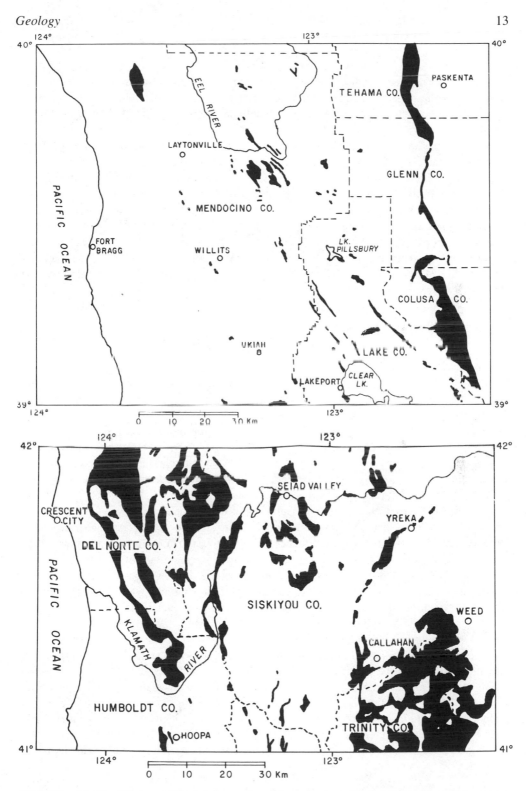

FIGS. 7 and 8. The Ukiah and Weed sheets, redrawn from Geologic Map of California (California Div. of Mines and Geology, 1958–1967) by Floyd Bardsley, showing major ultramafic outcrops (darkest areas) in the north Coast Ranges. 1 cm = 12 km.

Sierra Nevada

Ultramafics are discontinuous along a narrow belt from the southernmost Sierra to near the northern terminus of the range at Lake Almanor (figs. 9 and 10). All outcrops are cismontane and mostly at low elevations, more so in the south and increasing slightly in elevation northward.

Apart from the mafic (gabbro) outcrop in the Piute Mountains of Kern County, Tulare County contains the first substantial serpentinites. From there northward, every foothill county of the Sierras, from Tulare and Fresno to Plumas, contains serpentine (see fig. 11). Some of the most notable large outcrops are in the Coulterville–Bagby area of Mariposa County and on either side of the Feather River Canyon in Butte and Plumas counties.

All the area sheets of the Geologic Map of California are now in print. Those covering the Coast Ranges and the Sierra Nevada should be consulted for specific localities, pattern of outcrop, etc. Serpentinite is easily spotted by its map color—shocking purple; its map symbol is "ub." (I have provided a more extended account of the distribution of ultramafics in Appendix A.)

HISTORY OF THE GEOLOGIST'S INTEREST IN THE ULTRAMAFICS OF CALIFORNIA

Early geologists in California were aware of serpentinites, even if botanists were not. Perhaps the first recognition of serpentine is on a geological map of the San Francisco Bay area published in 1826 (figured in Norris and Webb, 1976, p. 275). This map, made by Edward Belcher, surveyor, and Alex Collie, surgeon, of H.M.S. Blossom, designated serpentine on the Tiburon peninsula and Angel Island. In the pioneer work on the geology of California (Whitney, 1865), the term "serpentine" was used several times in connection with outcrops visited by the Survey party. Yet in their visit to New Idria, the most spectacular serpentine barren in the West, the Survey writers seem not to have connected that vast tract of barren land with the serpentine substrate.

Only much later, in the early 20th century, did the connection between serpentine and vegetation appear in print. Two reports on quicksilver resources of California (Aubery, 1908; Bradley, 1918) make numerous references to serpentine, as it is so commonly associated with quicksilver ore deposits. The link between serpentine and plant life is finally made explicit in Bradley's 1918 quicksilver report; in describing the ore deposits and mines of the Mayacamas Mountains region of Lake County (p. 31), Bradley states: "Even where not bare, the serpentine can be detected at a distance by a sparse vegetation." In the same report, a photograph of the New Idria landscape (p. 94) is captioned "Serpentine surface near New Idria, San Benito County, showing characteristic sparseness of timber and brush growth." So at least field geologists had begun to notice the striking coincidences of serpentine rocks and the barrenness of the vegetation.

The paucity of early recorded observations from both botanists and geologists on serpentine and its link to the flora seems strange, especially in light of the fact that the serpentine phenomenon had long been known in Europe. Geological use of the term "serpentine" goes back before the 16th century (Challinor, 1967; Faust and Fahey, 1962); the allusion to "serpent" was to suggest the likeness between the mottled pattern of a serpent's skin and that of the rock.

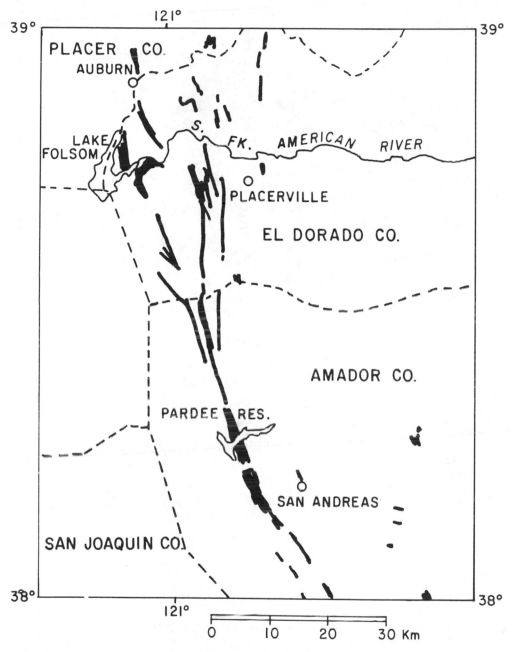

FIG 9. The Sacramento sheet, redrawn from Geologic Map of California (California Div. of Mines and Geology, 1958–1967) by Floyd Bardsley, showing representative ultramafic outcrops (dark areas) in the Sierra Nevada foothills from Amador to Placer counties. 1 cm = 7 km.

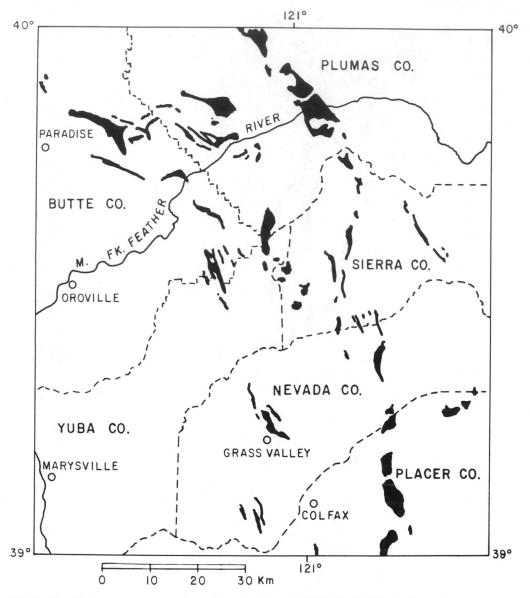

FIG 10. The Chico sheet, redrawn from Geologic Map of California (California Div. of Mines and Geology, 1958–1967) by Floyd Bardsley, showing representative ultramafic outcrops (dark areas) in the Sierra Nevada foothills from Placer to Plumas counties. 1 cm = 7 km.

An intriguing account of the history of western civilization's association with rocks and minerals of the serpentine group is given in Faust and Fahey's review (1962). The term *serpentine* was latinized from the Greek *lithos ophitidis*;Agricola in 1546 used the name "serpentinaria," a borrowing from the Greek of Dioscorides and Pliny. The Greek root *ophitidis* is often used both in geology and botany for attributes of ultramafic origin, and the term *ophiolite* has gained prominence in recent times as a collective term to embrace "a distinc-

FIG. 11. Digger pine (*Pinus sabiniana*) in association with mixed conifer forest on serpentine, Washington Ridge, Nevada County. (Photo by J. Griffin.)

tive assemblage of mafic to ultramafic rocks'' (Coleman, 1977, p. 6). In this sense it is neither a specific rock name nor a mapped lithological unit. This contemporary usage is usually framed in the context of plate tectonics.

The present term *serpentine* is found in variant form in many European languages. And its meanings are many: it has been used to refer to rock, soil, vegetation, and even in architecture and sculpture where ultramafic rock is used. At present geologists prefer the term *serpentinite* for rocks containing serpentine-group minerals. Though stricter usage may be preferred, I suspect we will continue to use the word ''serpentine'' to refer to the physical and biological attributes of ultramafic lithologies.

Botanists in Europe were writing about the plant response to serpentine in the mid-19th century. Amidei (1841) and Pancic (1859) are the earliest writers on the subject known to me; other 19th and early 20th century papers are cited in the reviews by Whittaker et al. (1954), Krause (1958), and Proctor and Woodell (1975). Given the precedence of European knowledge on plant life of serpentines, it is surprising that early scientific interest in the phenomenon of serpentine flora and vegetation in California did not surface in print in those early days of botanical activity in the state.

SERPENTINE SOILS AND THE MINERAL NUTRITION OF PLANTS

SERPENTINE SOILS

The effect of serpentines on plant life is elicited not so much directly by the rock itself as by its weathered product, serpentine soil. Soils weathered from ultramafic rocks strongly reflect the elemental composition of the parent rock (figs. 12 and 13). Soils rich in magnesium, iron, and silica are characteristic products of the weathering processes on ultramafic minerals. Yet the transformation from rock to soil is not a simple one; changes in mineral composition and ionic status usually occur. The clay fraction of a serpentine soil often consists of iron-rich montmorillonite, a mineral not found in the rock. Also the soil has less magnesium than does the parent rock. Wildman et al. (1968a,b) discuss these attributes of California soils derived from serpentinite. The depletion of magnesium during weathering is apparently caused by dissolution and leaching in a CO_2-rich aqueous environment. Montmorillonite is one of the characteristic clay colloids of serpentine soils in various parts of the world: besides their samples from California, Wildman et al. (1968b) analyzed soils from Borneo, Italy, and Japan, all containing montmorillonite. Soils with montmorillonite clays tend to be heavier in texture, retain moisture longer, and are more intractible to cultivation. The weathering of rock to soil does not always produce new clay minerals; rather, residual minerals of the parent rock form the colloidal fraction of the soil (Proctor and Woodell, 1975). Whether this occurs in any California serpentine soils is not known; probably both residual minerals and new clay minerals can be found.

Strictly speaking, a distinction should be made between soils derived from serpentinite (serpentinized ophiolites), just described, and peridotites. The latter parent materials characteristically form laterite soils; Jenny (1980, p. 183) gives as examples the red soils of northwestern California, e.g., at Red Mountain near Leggett.

Several soil series in California have been identified with ultramafic parent materials. USDA Soil Conservation Service Soil Surveys list the following soil series on serpentine parent materials:

> *Upland Series* (residual soils on ultramafic parent material): Henneke, Montara, Dubakella, Climara, Delpiedra, Fancher, and possibly Gilroy (on basic igneous). The Henneke and Dubakella Series are the most commonly mapped.
> *Basin (alluvial) Series*: Polebar, Venado, Maxwell and Conejo.

At least three soil series associated with ultramafic rock were sampled by Wildman et al (1968b) in their studies on clay minerals; two were from the Coast Ranges and were classified as Henneke and Dubakella gravelly loams. They also sampled a Fancher gravelly loam;

18

FIG. 12. Samples of serpentinite rock and its derived soil. (Samples from the north Coast Ranges).

this series occurs over serpentine rock in the Sierra foothills. Other upland soil types on serpentine have not yet been recognized by soil surveys—the serpentine soils at New Idria were not given a series name in the Wildman paper.

Most serpentine soils are residual or colluvial, formed in place over the parent rock. Such soils are shallow and stony; some have only A and C horizons; all are highly erodable. The upland serpentine soils typify the skeletal or lithosol group of soils; in the "Seventh Approximation" for soil taxonomy (USDA Soil Survey Staff, 1975) they would be called lithic argixerolls. Alluvial serpentine soils do occasionally occur, where their source materials are exclusively alluvial products from nearby residual serpentine upslope, as at New Idria and the Middletown area of Lake County. Alluvially derived serpentine soils can have a much deeper profile and a finer texture, and may have a substantial clay fraction. (A more detailed account of soil series derived from serpentine is given in Appendix B.)

EXCHANGEABLE ELEMENTS AND MINERAL NUTRITION OF SERPENTINE SOILS

The stressful and highly selective nature of serpentine habitats is undoubtedly a consequence of the interplay of physical, chemical and biotic factors. Having stated this ecological truism, it is still possible to single out one of several limiting factors to plant growth that appears to dominate plant response to the substrate. That factor is soil chemistry. Serpentine soil, the weathered product of ultramafic rock, has, like other soils, the capacity to provide inorganic nutrients and other elements from the soil solution and the clay colloid fraction of

FIG. 13. Azonal serpentine soil at the barrens of the Clear Creek-New Idria area, San Benito County. (Photo by J. Griffin.)

the soil. But it is the quality of this ion-exchange capacity and ionic reserve that distinguishes serpentine soils from those derived from nonferromagnesian rocks. The numerous chemical analyses of California serpentine soils (Robinson et al., 1935; Gordon and Lipman, 1926; Martin et al., 1953; Walker, 1954; Vlamis and Jenny, 1948; and Kruckeberg, 1954) all show a consistent pattern (Table 1): 1. Levels of exchangeable magnesium are much higher than any other cation; 2. calcium levels are usually lower than those found on nonserpentine soils; 3. levels of nitrogen, potassium and phosphorus are usually below those required for normal growth of crop plants; and 4. the heavy metals, chromium and nickel are often in high concentrations, while molybdenum may be in amounts insufficient for normal growth. The overall chemical infertility of California serpentines has been accepted as the controlling factor in plant response to serpentine soils (Proctor and Woodell, 1975). But which of these atypical chemical attributes plays the crucial role in promoting the "serpentine effect?"

Most serpentine soils in California range from slightly acid to moderately alkaline. Besides the pH values in Table 1, which show a range from 6.1 to 8.8, I summarize here pH values for 38 soil samples. These determinations were made on the air-dried, 2 mm fraction of samples of serpentine and nonserpentine soils in the immediate vicinity of species of *Streptanthus*, well-known serpentinicolous crucifers. The mean pH value for 23 nonserpentine soils was 6.3 (range 4.7–8.0); the mean value for 15 serpentine soils was 7.2 (range 5.6–7.8). However, the pH values for waters issuing from ultramafic rocks are much higher and appear to be bimodal (Barnes, et al., 1966). Waters of meteoric (= atmospheric) origin are moderately alkaline (pH range of 8.3–8.6 in waters charged with magnesium bicarbonate), while waters associated with contemporaneous serpentinization of pyroxene and olivine are strongly alkaline (pH range of 11.2–11.8 in waters charged with calcium hydroxide) In passing, I was much intrigued with the inference drawn by Barnes et al. (1966) from their data, that this highly alkaline water is geochemical evidence for present-day serpentinization.

Table 1. Base exchange capacity, exchangeable calcium and magnesium, and pH of some ultramafic soils, California and the Pacific Northwest[a]

BEC	Ca	Mg	Ca/Mg	pH	Sample no.	Locality
						California, Coast Ranges
17.2	4.1	11.1	0.4	7.0	167	Santa Barbara Co.
34.0[b]	52.5	2.7	19.4	7.5	150	San Luis Obispo Co.
15.0	2.4	11.4	0.2	7.2	138	San Benito Co.
9.0	1.5	5.2	0.3	7.2	127	Santa Clara Co.
5.9	0.5	3.9	0.1	7.5	168	Sonoma Co.
16.0	2.8	11.8	0.2	7.0	135	Lake Co.
43.0	5.0	32.5	0.2	8.8	134	Tehama Co.
17.0	2.8	12.7	0.2	6.7	129	Trinity Co.
14.0	2.8	7.1	0.4	7.1	152	Del Norte Co.
						California, Sierra Nevada
15.0	2.4	10.1	0.2	5.6	120	Butte Co.
11.0[b]	2.8	1.0	2.8	5.4	133	Plumas Co.
						Oregon
25.0	2.3	15.2	0.15	6.6	51	Grant Co.
						Washington
18.4	4.7	11.1	0.4	7.0	3	Kittitas Co.
14.2[b]	7.4	0.1	7.4	6.1	6	Kittitas Co.
19.9	3.9	14.8	0.3	6.2	69	Whatcom Co.
						British Columbia
—	3.4	7.4	0.5	6.9	82	Bralorne area

[a]Values in first three columns are expressed as millequivalents per 100 grams, 2 mm fraction, air-dry soil; all samples by the author.

[b]A nonserpentine soil for comparison.

PHYSIOLOGICAL AND MORPHOLOGICAL RESPONSES TO SERPENTINE

CATIONS

Over the years, each particular chemical factor has had its protagonist as *the* limiting component of serpentine. Gordon and Lipman (1926) viewed the deficiency of nitrogen and phosphorus as the major determinants of plant growth on serpentines. The exceptionally high concentrations of magnesium, presumably inducing magnesium toxicity, have been singled out as critical, mostly in Europe (Novak, 1928; Proctor, 1970). Because of the critical role of calcium in the mineral nutrition of plants, its very low levels in serpentine soils have been thought both to cause the exclusion of many species and to have induced evolutionary adaptive responses in serpentine taxa (Kruckeberg, 1954, 1969a; Walker, 1954; Vlamis and Jenny, 1948; Vlamis, 1949). Since calcium and magnesium co-vary in soils, and fertile soils have a high calcium-to-magnesium ratio, some investigators have seen the low Ca/Mg ratio (<1.0) of serpentine soils as playing a critical and selective role in plant growth (Madhok and Walker, 1969; Walker, 1954; Walker et al., 1955; C.D. White, 1971).

In a recent study of Koenigs, et al. (1982), some serpentine species in eastern Napa County were analyzed for foliage content of major nutrient elements (K, Ca, Mg). As with other serpentine species (Walker, 1954) the authors found that *Cupressus sargentii* accumulated Ca and excluded Mg. "*Adenostoma fasciculatum* occurred mostly on soils with a relatively high Ca content, had the lowest Ca concentration in its leaves, was most abundant where Ca concentration in its leaves was highest, and had the highest Mg accumulation ratio." The authors suggest that *A. fasciculatum* is the least adapted of the species studied to serpentine habitats.

HEAVY METALS

Those who see mineral toxicity as a primary cause of exclusion (and of physiological adaptation) have ample justification for singling out chromium and nickel. Both these heavy metals are often highly concentrated in serpentine soils (Proctor and Woodell, 1975). Of the two heavy metals, nickel is more likely to affect plant growth; the biological effects of chromium are lessened, since chromium is usually in an insoluble form in serpentine soils (Proctor and Woodell, 1975). Although there is a substantial literature on the effects of these two heavy metals on serpentine floras in other parts of the world (e.g., Europe, New Zealand, New Caledonia, and Rhodesia), there are scarcely any published reports on the interaction of chromium and nickel on the serpentine floras of California. Since the identification of nickel "hyperaccumulator" plants in other regions (in New Caledonia, Jaffre et al., 1976,

1979; in Rhodesia, Wild, 1970), a similar phenomenon was to be expected in California. (Hyperaccumulator species are plants that accumulate more than 1000 ppm of a heavy metal in their tissue.) One such taxon has now been identified in the serpentine flora of California: in a survey of species of *Streptanthus*, crucifers well known for their high fidelity to serpentine (Kruckeberg 1958, 1969a; Morrison, 1941), only *S. polygaloides*, a Sierra foothill serpentine endemic, fits the character of a hyperaccumulator (Reeves et al., 1981). Nickel values for several collections of *S. polygaloides* ranged from 1100 to 16,400 ppm.

Deficiency in the essential micronutrient element molybdenum is known to occur in at least some California serpentines (Walker 1948), but low molybdenum has not been invoked as the critical "serpentine factor." R.G. Coleman (personal communication.) remarks that "analyses of serpentinites and peridotites show molybdenum to be below detection levels."

THE SERPENTINE SYNDROME

It is not surprising that single chemical constituents, appearing in some extreme form—toxic or deficient—have been seized upon to account for the serpentine phenomenon. The single-factor approach, if true, would be so simple! But ecosystems function as multifactorial, holocoenotic systems; single-factor dependence is merely a simplifying device for the laboratory experimenter. In recent years the causal basis of serpentine vegetation and floristics has been studied from the more realistic, multifactorial position (Proctor and Woodell, 1975). The ensemble of chemical, physical, and biotic factors, with their particular intensities, forms the feedback loop that orchestrates the "serpentine syndrome" (Jenny, 1980).

The network of operational factors that yields a serpentine vegetation and flora can be diagrammed (fig. 14) for heuristic value. The holocoenotic view implies that plants must respond genotypically to the complex of factors in the serpentine habitat with an array of structural and functional attributes. The exceptional chemical composition of serpentine soil—low calcium, high magnesium, heavy metals, etc.—sets in motion a biological response that results in a low turnover of nitrogen and phosphorus. The low nutrient status and the cation imbalances in turn promote a sparse plant cover, and thus a high heat budget may result; high temperature effects, moisture stress, and biotic effects (Tadros, 1957) may then further check plant growth and survival. Since serpentine soils are usually residual soils, derived from rock outcrops on steep or irregular topography, the habitats are often unstable talus—adding a further stressful challenge to plant occupancy. Evolutionary accommodation to this suite of stresses has taken two basic forms: 1. the genesis of species uniquely restricted to serpentine; 2. racial differentiation of wide-ranging species to yield serpentine-tolerant ecotypes. Extensions of distributional ranges in altitude, latitude, and new habitat type, as well as sparse plant cover and low yield of biomass for serpentine vegetation, are common ecogeographic responses.

MORPHOLOGICAL RESPONSES

Morphological characteristics of serpentine flora— both known and anticipated—include: 1. xeromorphic foliage (sclerophylly, glaucousness, size reduction, reduced or increased pubescence, anthocyanous coloration, etc.); 2. reduction in stature (shrubbiness of arborescent species, dwarfing and plagiotropism of herbaceous species); and 3. increase in root system (Pichi-Sermolli, 1948; Ritter-Studnika, 1968; Krause, 1958). The overall reduction in

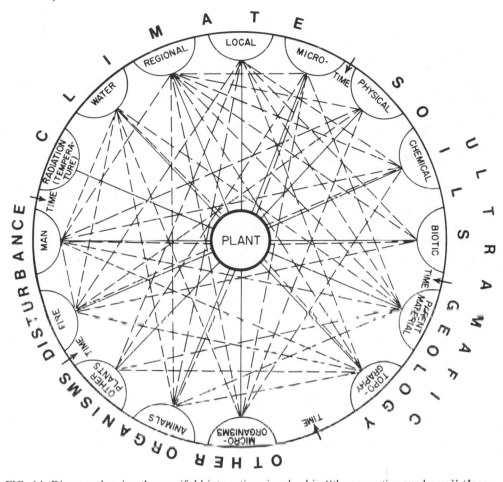

FIG. 14. Diagram showing the manifold interactions involved in "the serpentine syndrome" (Jenny, 1980). (Diagram modified from Billings, 1952.)

growth form of either the whole plant or its parts may be a result of slower growth rates for serpentinicolous taxa. Parsons (1968) has shown that slower growth rates are genetically fixed in species of stressful edaphic habitats.

OTHER RESPONSES

Ecophysiological adaptation to serpentine has to reckon with several chemical and physical stresses, acting in concert. Plants can respond in more than one way to the high magnesium levels: the cation may be excluded or sequestered in a nonliving cell wall of roots, be restricted in its uptake beyond the roots, or be actively excluded from any tissue (Peterson, 1971; Proctor and Woodell, 1975). Accommodation to low calcium levels usually involves preferential accumulation of that essential element (Walker, 1954). Response to low soil nitrogen and phosphorus levels may include selective absorption of the low amounts available (Proctor and Woodell, 1975). Response to heavy metals, nickel and chromium, most often involves a mechanism of exclusion or rendering the cations innocuous (Antonovics et al., 1971).

Inherited adjustment to the several physical stresses— high temperature, low moisture, and soil instability—has scarcely been examined. The changed morphology of serpentine races or species can often be taken as an indication of genotypic response to physical stresses. In a word, genetic adaptation to serpentine is surely compounded of a complex of attributes. It should be remembered that this estimation of the complexity of plant adaptation to serpentine is based on information about serpentine plant life at many different places around the world. Though I strongly suspect that the serpentine phenotype in complex manifestation has been assumed by plants on the serpentines of California, there remains a great hiatus in our understanding of genecological/ecophysiological traits associated with California serpentine plants.

Yet another variable in the physiological response of plants to serpentine is the set of species-to-species differences in the ways plants cope with the elements in serpentine soils. Though next to nothing is known about this aspect for California, investigators elsewhere have demonstrated that species within a given serpentine flora, cope differently, one from another, with the elements encountered in the soil—e.g., hyperaccumulators. Physiological adaptation can take the form of exclusion, avid uptake, or sequestering in innocuous form one or more of the characteristic elements of serpentine soils. For instance, Lyon et al. (1971) found that one species excluded magnesium, another preferentially accumulated calcium, and yet another was able to exclude chromium and nickel. Their conclusion, "that there was no single universal mechanism to explain plant survival on serpentine soils of New Zealand," probably applies to plants of California serpentines and elsewhere in the world. Such a highly individualistic response among serpentinicolous taxa is borne out further by the hyperaccumulator effect cited earlier: although several species of *Streptanthus* from serpentine substrates in California were sampled, only one, *S. polygaloides*, accumulated large amounts of nickel (Reeves et al., 1981).

SERPENTINE VEGETATION IN CALIFORNIA

Serpentines intrude upon the California landscape mostly at moderate elevations and wholly west of the Sierra Nevada crest. Thus it is grassland, chaparral, oak woodland, and conifer forest that must meet the challenge of serpentine substrates. Moreover, since most serpentine outcrops are in a xeric, Mediterranean climate, the added stress of drought to chemical imbalance makes for even more severe environments. The net effect on plant life is the striking discontinuity between vegetation on the nonserpentine bedrock supporting normal vegetation and the plant cover on serpentine.

The serpentine barren is the most extreme manifestation of the effect of serpentine parent material. Some barrens have no plant cover whatsoever; more often the barren supports a sparse herbaceous or shrub/herbaceous cover with much intervening barren ground. The contact between normal and serpentine soils can be so abrupt that one imagines the vegetation on the serpentine side to have been severely burned or grazed. Yet this sharp contrast was not contrived by man, but by natural sorting of a select few plants tolerant to serpentine and exclusion of much of the adjacent flora. Contact zones are often sharp, and the species composition, density, and pattern of distribution in communities on either side of the contact are markedly distinct. Often the contrast in vegetation takes the form of a sudden shift in life-form spectrum and physiognomy. The most frequent shifts are from blue oak and digger pine to hard chaparral or a sparse grass–forb cover, or from chamise chaparral to a – scrub. Whittaker (1975) comments on this shift in life-form spectra as a worldwide phenomenon: in Quebec, the shift is from taiga to tundra; in Oregon, it is an abrupt transition from Douglas fir to open pine woodland; in California, it is from oak woodland to chaparral; in Cuba and New Caledonia, the shift can be dramatic, from tropical forest to savannah-scrub; and in New Zealand, it is from southern beech forest to tussock grassland.

Practically every major vegetation type at low to mid elevations from Santa Barbara County to the Oregon border is touched by the serpentine effect. Nearly always the serpentine–nonserpentine discontinuity occurs on residual soils, upslope from alluvial formations. The distribution of serpentine habitats can be overlain on a vegetation map of California (Kuchler, 1977) to reveal the many different vegetation types that can border serpentine. Besides the most common ones just mentioned, serpentine "islands" occur in redwood forest, yellow pine forest, coastal cypress–pine forest, Klamath and Coast Range montane forests, mixed evergreen forest, mixed hardwood forest, and California prairie and coastal sagebrush.

Discontinuity of vegetation type and floristic composition usually results when any of these vegetation zones is intruded by serpentine. The Kuchler map does not recognize these incursions of serpentine in the vegetation landscape, partly because of scale. Undoubtedly, though, particular patterns within the intricate mosaic of different vegetation formations in the north and south Coast Ranges do reflect the influence of the larger serpentine areas as they abut other formations. Chaparral–woodland contacts are likely to be serpentine–nonserpentine contacts; but since the variant floristic composition of chaparral is not differentiated, contacts due to serpentine are thus obscured in this map. The only formation on serpentine that Kuchler recognizes is the very restricted San Benito Forest (*Calocedrus decurrens* (Torr.) Florin) near New Idria.

Any more fine-grained view of cismontane California vegetation is bound to include communities of species peculiar to serpentine. Serpentine chaparral (Hanes, 1977, pp. 429-430) and serpentine grassland (Howell, 1970) are recognized as unique vegetation formations. Scattered through Barbour and Major (1977) are references to unusual range extensions, conifer distributions, plant groupings, and vegetation discontinuities on serpentines.

The characteristics of serpentine chaparral are nicely summarized by Hanes (1977 pp. 429–430):

> Serpentine chaparral is an open, low type associated with serpentine soils from San Luis Obispo Co. northward through the Coast Ranges and foothills of the northern Sierra Nevada. The shrubs are characterized by apparent "xeromorphism" (peinomorphism, serpentinomorphism) and dwarfed stature resulting from reduced productivity and growth (Whittaker, 1954). The dominant shrubs are *Adenostoma fasciculatum* H. and A. and *Heteromeles arbutifolia* M. Roem., but noteworthy are several localized endemic shrub species, *Arctostaphylos viscida* and *Ceanothus jepsonii*.
>
> *Cupressus sargentii, Garrya congdonii*, and *Quercus durata* are unmistakable "indicator species" because of their typical restriction to, and numerical dominance on, serpentine soils (Kruckeberg, 1954). Serpentine chaparral may be associated with foothill woodland (*Pinus sabiniana Dougl.*) or montane coniferous forest (*P. jeffreyi* Grev. and Balf., *P. ponderosa* Dougl. ex P. and C. Lawson, *P. attenuata* Lemmon, *Pseudotsuga menziesii* (Mirbel) Franco) as an understory (Whittaker, 1954). The thousands of hectares of serpentine chaparral in the north Coast Ranges are easily distinguished from the oak–grasslands on hills of nonserpentine origin (Walker, 1954). Wells (1962) has described serpentine chaparral in San Luis Obispo Co., as has Vogl (1973) for the Santa Ana Mts., Orange Co.

(See figs. 15–19 for views of serpentine chaparral in the Coast Ranges.)

In his flora of Marin County, Howell (1970) recognizes two serpentine communities. The chaparral–Sargent cypress association on Mt. Tamalpais consists of several woody serpentine species, *Quercus durata, Ceanothus jepsonii, Arctostaphylos montana* Eastw., and the more arborescent *Cupressus sargentii*, along with many grasses and other herbs, some of which are obligate serpentine species. Serpentine grassland in Marin County is restricted to the Tiburon peninsula and includes some of the most restricted endemics of California: *Strepthanthus niger, Castilleja neglecta* Zeile in Jeps., and *Calochortus tiburonensis.* Serpentine grassland is much more local than serpentine chaparral, and (without the Tiburon endemics) is to be expected elsewhere in coastal areas such as the Redwood, California Prairie, and Coastal Sagebrush associations. Annual fescues (*Festuca pacifica* Piper and *F. reflexa* Buckl), *Stipa pulchra* Hitchc., as well as naturalized Europeans like *Bromus rigidus* Roth., *B. mollis* L., and *Avena fatua* L., are the chief grass species of this formation.

McNaughton (1968) has made some revealing comparisons between serpentine and nonserpentine grasslands; his sample of the California grassland type was in the Jasper Ridge Reserve of Stanford University, in San Mateo County—the same site used by Cooper (1922) in his classic study of California chaparral. McNaughton found that the nonserpentine grassland at Jasper Ridge, which occurs on sandstone, had greater productivity and a higher annual yield of biomass, but that the serpentine grassland showed greater species diversity. The biological attributes of the sandstone grassland were correlated with the habitat gradient (cool, moist to warm, dry); no such relationship was found for the serpentine grassland.

FIG. 15. Serpentine chaparral in the Clear Creek area, San Benito County. (Photo by J. Griffin.)

It is evident that California botanists have given more attention to the floristics of serpentine than to its effect on the synecology or autecology of the flora. I am aware of only a few other ecological studies of serpentine vegetation. Whittaker's (1960) study of the vegetation of the Siskiyou Mountains is a landmark paper, for here the concept of gradient analysis was successfully applied to vegetation on contrasting substrates. Though Whittaker's study sites were mainly in Josephine County, Oregon, the same vegetation types continue into northwestern California. Low-elevation serpentine vegetation occurs along the moisture gradient from open *Chamaecyparis lawsoniana* (A. Murr.) Parl.–*Pinus monticola* Dougl.–*Psuedotsuga* mesic forest, "through very distinctive forest–shrub stands with several conifers and a two-phase undergrowth of sclerophyll shrubs and grass to the [more xeric] *Pinus jeffreyi* woodlands."

In the same region as Whittaker's study, Zobel and Hawk (1980) have examined the autecology of *Chamaecyparis lawsoniana* (Port Orford cedar). This conifer is confined to wet ultramafics except at high altitudes and northern localities (in Oregon). It seems to be confined to sites of low moisture stress (minus 11 bars at predawn); certainly the high frequency of seeps and springy areas on ultramafics in the Siskiyou region of Oregon and California can account for Port Orford cedar's local fidelity to ultramafics. Though Zobel and Hawk do not comment on the chemical aspects of the substrates, it is not unlikely that Port Orford cedar has accommodated to the exceptional base status and nutrient levels of ultramafics while taking advantage of the mesic conditions. Zobel attributes the success of Port Orford cedar

FIG. 16. Serpentine chaparral-cypress woodland at Carson Ridge, north slope of Mt. Tamalpais, Marin County.

on ultramafics to the wetness of the sites and the absence of competition from Douglas fir and other woody species.

South of the more mesic-mixed conifer vegetation on serpentine of the Klamath–Siskiyou country, vegetation in general—and particularly that on serpentine—shows the effect of increasing drought stress (both apparent and real—see below). Only a few ecological studies of this more xeric-appearing vegetation are on record. Wells (1962) found great diversity in vegetation pattern and floristic composition from his study of the vegetation on a variety of substrates (more than 15) in the San Luis Obispo region. Wells contended that serpentinite rock causes a change in the flora rather than a shift in vegetation type. Moreover, disturbance, especially fire, probably accounts for the physiognomic differences both within and between substrates. Wells cautions against the superficial inference that serpentine sites are xeric; despite xeric taxa like *Yucca, Quercus durata*, and other xeromorphic plants, the accommodations of *Umbellularia* and other more mesic and even riparian taxa suggest that serpentine sites in his study area may be more moist than is supposed. This idea is amplified by Hardham (1962) in her observations on the floristics of serpentine in the outer south Coast Ranges.

The vegetation on the only known locality of serpentine in southern California—in the Santa Ana Mountains—was studied by Vogl (1973). The hydrothermally altered serpentinite

FIG. 17. Serpentine chaparral on the Paskenta-Covelo road, western Tehama County. *Quercus durata* and *Ceanothus jepsonii* are codominant shrubs.

supports a locally restricted stand of knobcone pine (*Pinus attenuata*). Other notable taxa restricted to this local ultramafic outcrop are *Ceanothus papillosus* T. and G. var. *roweanus* McMinn and *Ribes malvaceum* Sm. var. *viridifolium* Abrams. The pine appears to be restricted to serpentinite here, due to the lack of competition with the chaparral species common on adjacent nonserpentine habitats as well as its adaptation to low nutrient levels. Vogl further supports the increasing realization that ultramafic soils can have a higher water-holding capacity than nearby nonserpentine soils: ''The water-retaining capacity of the serpentinite is nearly double that of the chaparral soils. This soil characteristic, the frequent fogs, the location of the pines in fog gaps, . . . contribute to the persistence of the pines. . . .'' The knobcone pine habitat is held in a successional state by fire, to which the closed-cone attribute is uniquely adapted.

Californian species of *Cupressus* and of the closed-cone pine group are characteristically restricted to serpentinite and other edaphically atypical soils. McMillan (1956) did extensive field and greenhouse studies of these coniferous species as well as of a number of associated herbaceous taxa. Since low calcium and reduced nitrogen and phosphorus levels were com-

FIG. 18. Serpentine chaparral landscape on the Paskenta-Covelo road, western Tehama County.

mon to both the serpentine and acid soils, McMillan expected that reciprocal progeny testing would show a "common denominator" effect: that is, a given strain of conifer species would be tolerant to acid *and* serpentine soils. Instead, edaphic tolerance appeared to express itself in a highly individualistic manner. However, tolerance to both acid and serpentine soils seems to occur in *Rhododendron occidentale* (T. and G.) Gray. It has been proposed that this azalea can thrive where the percentage of exchangeable calcium is low—as it is on both serpentine and the acid soils in California where the plant "normally" occurs (Leiser, 1957).

Principal components analysis of data from serpentine vegetation in eastern Napa County (Koenigs et al., 1982) achieved a grouping of "cypress" and "non-cypress" stands. "Stands with *Cupressus sargentii* also contained *Arctostaphylos viscida* and occurred on mesic sites with lower Ca in the sub-surface soil. Stands without *C. sargentii* usually contained *Adenostoma fasciculatum* and *Quercus durata* and occurred on the drier sites with higher Ca."

FIG. 19. Serpentine chaparral with scattered digger pine (*Pinus sabiniana*) in a mixed conifer forest setting, north fork of Beegum Creek area, Trinity County.

The abrupt discontinuity between serpentinite and other substrates often evokes a sharp break in plant communities—both in life-form and in species composition. The only ecological analysis of such a contact discontinuity for Californian vegetation is in the paper by J.T. Gray (1979), in which the vegetation along elevational gradients in the north Coast Ranges and the Sierra was studied. In his work on the Coast Range sequence at Snow Mountain (Lake County), Gray noted an abrupt change from chaparral woodland on lower-elevation serpentinites to montane coniferous forest on higher-elevation nonserpentine parent materials. The attributes of a clear-cut break in vegetation type are mainly floristic, but the shift in taxa also includes a shift in life-form. Along the Snow Mountain transect, Gray found the common woody plants of serpentinite (*Pinus sabiniana, Ceanothus cuneatus* (Hook.) Nutt., *Quercus durata* and *Adenostoma fasciculatum*) giving way to *Pinus ponderosa, Quercus chrysolepis* Liebm., *Arctostaphylos canescens* Eastw., *Pinus lambertiana* Dougl., and *Abies concolor* (Gord. and Glend.) Lindl. on other substrates. The sharp break in vegetation type at 1250 m coincides with the change from ultramafic to normal soils.

SERPENTINE FLORA OF CALIFORNIA

The challenges of the serpentine environment have stimulated a variety of responses in the flora of California. 1. The most spectacular response has been the evolution of species that are narrowly restricted to serpentine—the "narrow edaphic endemics" of Mason (1946a,b). 2. Many species not restricted to serpentine may, however, appear on serpentine at localities distant from their normal ranges of distribution; these so-called local indicators (Peterson, 1971), when they involve dominant trees or shrubs, may make a sharp contrast to adjacent nonserpentine plant life. 3. A more subtle response is ecotypic variation, where a wide-ranging species has evolved races or populations genetically tolerant to serpentine (Kruckeberg 1951, 1954, 1967); this parallels the climatic race response so elegantly demonstrated by Clausen, Keck, and Heisey (1940, 1948). 4. An obvious yet overlooked response is that of exclusion: many species on adjacent nonserpentine sites do not appear on serpentine. We now examine each of these responses in more detail. Pictures of some serpentine plants are grouped in figs. 20–32 (Appendix F).

SERPENTINE ENDEMISM

No other serpentine occurrences in the temperate zones of the world have so rich an endemic component as the serpentines of California. From a variety of sources (published and unpublished records, herbarium searches, and local and regional floras), I reckon that 215 vascular plant taxa (species and infraspecific variants) are either wholly or largely restricted to serpentines in California.[1]

Tables 2 and 3 highlight some of the statistics on serpentine endemism. Some caution should be used in evaluating these figures, as much of the data was taken from secondary sources. For example, the two California floras (Munz and Keck, 1959; Abrams, 1923–1960) often mention serpentine restriction in the diagnoses of particular taxa; not all these endemics have been confirmed by me or by other field botanists. Moreover, it is very probable that these two floras have omitted a number of serpentine taxa. Of the many additions to the California flora since the publication of these two major compendiums, some taxa are serpentine endemics; newly discovered serpentinicolous plants come to light almost yearly. This is particularly true of the little-explored serpentines of northwestern California (the Klamath–Siskiyou Mountains). One further caveat: just as there are no absolutes in other fields of biology, so it is for serpentine endemism. There are instances where a taxon is almost wholly restricted to serpentine, but may occasionally occur on a nonserpentine substrate. This seems more often to be the case for wide-ranging taxa: thus *Quercus durata*, though essentially restricted to serpentine, has an occasional nonserpentine locality, though

---

1 A full listing of taxa endemic to California serpentines is given in Appendix C.

34

in this case the nonserpentine stations may have introgressant populations—*Q. durata* X *Q. dumosa* Nutt. (thanks to John Tucker, personal communication).

Restrictions to serpentine may be very local—a single outcrop—to regional and widespread occurrences. Species of the cruciferous genus *Streptanthus* illustrate each of these three levels of endemism: *S. niger* of the Tiburon peninsula, *S. batrachopus* of Mt. Tamalpais, and *S. brachiatus* of the Mayacamas Mountains are striking examples of very local, distinctive endemics; the intermediate level, where a taxon is restricted to serpentine outcrops of a broader regional area, is represented by *S. hesperidis* Jeps. (Napa and Lake counties) and *S. polygaloides* (from Fresno to Butte counties on serpentines of the Sierra Nevada foothills); and *S. glandulosus* Hook. can be found on Coast Range serpentines all the way from San Luis Obispo County to Josephine County in Oregon, nearly the full extent of the Franciscan ultramafics in California. Probably the most widespread of all the strict serpentine endemics is *Quercus durata*: which occurs throughout the Coast Ranges from San Luis Obispo County to Trinity County and in the Sierra from Nevada County to Eldorado County.

Serpentine endemism can occur in all major vascular-plant groups: conifers, ferns, monocots, and dicots. Though the largest number are in the dicots, they are unevenly distributed among families; Table 3 gives a numerical summary of frequency of endemism at the family level.

Particular genera within these dicot families have spawned more than one or two serpentine endemics: *Streptanthus* (Cruciferae), 16 taxa; *Eriogonum* (Polygonaceae), 10 taxa; Arabis (Cruciferae), 5 taxa; *Astragalus* (Leguminosae), 3 taxa; *Linum* (=*Hesperolinon*) (Linaceae), 8 taxa; *Ceanothus* (Rhamnaceae), 4 taxa; *Lomatium* (Umbelliferae), 8 taxa; *Arctostaphylos* (Ericaceae), 4 taxa; *Phacelia* (Hydrophyllaceae), 5 taxa; *Monardella* (Labiatae), 5 taxa; *Castilleja* (Scrophulariaceae), 3 taxa; and *Cordylanthus* (Scrophulariaceae), 5 taxa

Several genera of the Compositae have more than the occasional serpentine endemic: *Cirsium*, 5 taxa; *Haplopappus*, 2 taxa; *Eriophyllum*, 2 taxa; *Senecio*, 5 taxa; and tarweed genera like *Calycadenia, Hemizonia, Layia,* and *Madia,* with 2 or more endemics per genus.

Of the monocot families with some predilection for serpentine, Liliaceae and Amaryllidaceae are the most prominent. Liliaceous genera with several serpentine endemics are *Calochortus*, with 7 taxa; *Erythronium,* 4 taxa; and *Fritillaria,* ca. 4 taxa. In the Amaryllidaceae, *Allium* has 7 endemic taxa. No grass or sedge genera have evolved a significant number of serpentine endemics, though there are a number scattered throughout the two families. On closer study, I would expect to see still more endemics discovered in those graminoid genera with a large number of native species: e.g., *Festuca* (section Vulpia), *Poa, Melica,* and *Carex.*

LOCAL OR REGIONAL INDICATORS NOT RESTRICTED TO SERPENTINE[2]

While endemism is the most spectacular floristic response to serpentine, another manifestation of serpentinicolous affinity is often nearly as striking. This phenomenon involves taxa that grow on both serpentine and nonserpentine soils, but do so in unique ways and with distinctive patterns of distribution. A phytogeographic caveat is needed here, however: Cate-

---

2   A full listing of taxa in this category is given in Appendix D.

Table 2. Vascular plant taxa[a] associated with California serpentines
[endemics = 215; local and regional indicators = 221; bodenvag
(indifferent) = 1108]

Region	Number	Examples
A. Serpentine-endemic taxa		
Sierra Nevada	13	*Allium sanbornii, Streptanthus polygaloides*
Sierra Nevada + Coast Ranges	13	*Garrya congdonii, Helianthus exilis*
Coast Ranges (inclusive)	192	*Cupressus sargentii, Allium falcifolium*
South Coast Ranges	40	*Streptanthus insignis, Calochortus obispoensis*
Bay Area	19	*Streptanthus niger, Clarkia franciscana*
Napa, Sonoma, Lake counties	27	*Streptanthus brachiatus, Senecio clevelandii*
North Coast Ranges	14	*Allium hoffmanii, Asclepias solanoana*
Klamath-Siskiyou area	19	*Phacelia dalesiana, Streptanthus barbatus*
NW California-SW Oregon	30	*Fritillaria glauca, Lilium bolanderi*
NW California-Pacific NW	2	*Polystichum lemmonii*
B. Serpentine indicators		
Sierra Nevada	14	*Orthocarpus lacerus, Clarkia arcuata*
Sierra Nevada + Coast Range	42	*Aspidotis densa*
Coast Ranges (inclusive)	152	*Streptanthus glandulosus, Pinus jeffreyi*
South Coast Ranges	28	*Eriogonum covilleanum, Lewisia rediviva*
Bay Area	7	*Streptanthus albidus, Eriogonum caninum*
Napa, Sonoma, Lake counties	10	*Calochortus vestae, Mimulus brachiatus*
North Coast Ranges	13	*Arabis subpinnatifida, Calocedrus decurrens*
Klamath-Siskiyou area	20	*Picea breweriana, Pinus balfouriana*

Table 2. (cont.)

NW California-SW Oregon	17	*Chamaecyparis lawsoniana, Poa piperi*
NW California-Pacific NW	2	*Pyrola picta dentata, Campanula scabrella*

C. Bodenvag (indifferent) taxa

Sierra Nevada	20	*Calycadenia truncata, Arctostaphylos mewukka*
Sierra Nevada + Coast Ranges	68	*Pinus sabiniana, Adenostoma fasciculatum*
Coast Ranges (inclusive)	45	*Dentaria californica, Fraxinus dipetala*
South Coast Ranges	22	*Arctostaphylos pilosula, Yucca whippleyi percursa*
North Coast Ranges	37	*Draba howellii, Abies magnifica shastensis*
Bay Area	3	*Ceanothus masonii, Arctostaphylos sensitiva*
NW California-SW Oregon	18	*Phlox adsurgens, Arctostaphylos nevadensis*
Pacific states	386	*Pinus monticola, Achillea lanulosa*
California (general)	509	*Fritillaria* spp., *Calochortus* spp.

[a]Ferns, conifers, and flowering plants, species and infraspecific variants.

gorizing such distribution patterns becomes somewhat arbitrary and artificial, since there can be a number of permutations of the attributes and one category may grade into another. Given this proviso, I detect two such categories of substrate fidelity.

The first type involves taxa whose main range of distribution is on nonserpentine substrates but which are clearly faithful to serpentine in areas beyond the main range. This peripheral or outlier distribution may express itself in extensions of range in altitude and/or to other physiographic or vegetation provinces. Two conifers, Jeffrey pine (*Pinus jeffreyi*) and incense cedar (*Calocedrus decurrens*), nicely portray features of this pattern.

The distribution of Jeffrey pine is centered in the Sierra Nevada, mainly on the dry eastern slope where it may form pure stands. Here it is widespread at moderate to high elevations (6000–9000 ft., 1830–2745 m), growing mostly on dry decomposed granites. However, in northwestern California and southwestern Oregon Jeffrey pine becomes a faithful serpentine indicator at low (3000 ft., 915 m) to moderately high elevations (Griffin and Critchfield, 1972). Whittaker's (1960) study sites in southwestern Oregon and adjacent California typify this high fidelity to serpentine and peridotite; on nearby diorite or olivine gabbro, yellow pine (*Pinus ponderosa*) replaces Jeffrey pine.

Table 3. Vascular plants associated with California serpentines, by taxonomic group (species and infraspecific taxa)

Taxonomic group[a]		Notable Plant Families[a]	
A. Serpentine-endemic taxa			
Ferns	2/0	Compositae	24/7
Conifers	2/1	Liliaceae (sensu lato)	19/7
Monocots	22/7	Cruciferae	17/6
Dicots	126/56	Polygonaceae	13/0
		Umbelliferae	11/3
		Linaceae	8/0
		Scrophulariaceae	7/3
		Hydrophyllaceae	5/1
B. Serpentine-indicator taxa			
Ferns	4/1	Liliaceae (sensu lato)	19/2
Conifers	9/1	Compositae	11/17
Monocots	28/3	Scrophulariaceae	12/4
Dicots	105/58	Ericaceae (Arctostaphylos)	16/1?
		Boraginaceae	9/0
		Ranunculaceae	8/5
		Hydrophyllaceae	6/0
		Umbelliferae	6/0
		Linaceae	6/0
		Portulaceaceae	6/3
C. Bodenvag (indifferent) taxa (genus/species/variant)			
Ferns	1/3/0	Compositae	0/188/10
Conifers	0/10/0	Gramineae	0/117/1
Monocots	6/345/2	Cyperaceae	0/98/0
Dicots	7/748/64	Liliaceae (sensu lato)	6/81/0
		Leguminosae	0/75/4
		Scrophulariaceae	0/36/0

---

[a]Numbers of taxa by species/infraspecific taxa; e.g., 10/3.

Jeffrey pine makes its only appearance as a native in the south Coast Ranges on the spectacular serpentine barrens of the New Idria region (San Benito Mountain, elev. 1597 m); accounts of this colony and the one at Pine Ridge in the Santa Lucia Mountains are given in Griffin (1975) and B. Zobel (1952). However, other occurrences in the Santa Lucia Mountains are not on serpentine (Griffin, 1975).

Incense cedar departs from its main range along the cordilleran axis (Sierra Nevada–Cascade ranges) to become locally common in many places throughout the Coast Ranges. While not exclusively on serpentine in the Coast Ranges, it can be considered a local or regional indicator for serpentine in many habitats where chaparral or mixed evergreen forest would reign–if the substrate were not serpentine. In the north Coast Ranges and in the Klamath–Siskiyou country, serpentine occurrences of *Calocedrus* are common. Good examples are localities along St. Helena Creek (Highway 29) in Napa County; Bartlett Springs in Lake County, and the lower slopes of Mt. Eddy in Shasta and Trinity counties. Again, there are occurrences on nonserpentine in northwestern California; but here, where serpentines are extensive, *Calocedrus* is a faithful serpentine indicator.

The appearance of coniferous species in isolated stands on abnormal substrates has caught the eye of botanists elsewhere in the western United States. Billings (1950) describes the curious stands of Jeffrey pine and ponderosa pine on hydrothermally altered andesite rock in Nevada—curious because these "islands" are completely surrounded by a regionally dominant sagebrush vegetation (*Artemisia tridentata* Nutt. and associated species). Bristlecone pine (*Pinus aristata* Engelm. = *P. longaeva* Bailey—Bailey, 1970) on the dolomites of the highest White Mountains of eastern California (Wright and Mooney, 1965) is another striking case of a specialized substrate inducing a profound vegetation and floristic change. These and many other examples of anomalous occurrences of conifers on exceptional substrates nicely parallel the similar instances for serpentine that I have referred to above. Gankin and Major (1964) see a common cause for these and countless other cases of disjunct or "extralimital" (and endemic) occurrences of taxa on special substrates. They invoke competition as the prime determinant; what provides the competitive differential is some attribute of the substrate—physical or chemical.

Two shrubs show parallels to the Jeffrey pine case. *Quercus vaccinifolia* Kell. is common on granites at high elevations in the Sierra Nevada, but in the Klamath–Siskiyou area it is most often on ultramafics, and at low to moderate elevations. *Arctostaphylos nevadensis* Gray probably has a similar pattern.

Local or regional fidelity of herbs to serpentine does occur, but the distribution patterns are more complex; nearly every taxon has to be dealt with as a separate, unique floristic problem. For instance, *Aspidotis densa* (Brackenr.) Lellinger and *Adiantum pedatum* L. var. *aleuticum* Rupr. are two ferns common on serpentine, the former in xeric and the latter in mesic to hygric habitats. Nearly every low- to mid-elevation serpentine outcrop has the *Aspidotis* as a character-species. Yet throughout its range in California and elsewhere, this fern occasionally occurs on nonserpentine soils (e.g., granitics in the Sierra, volcanics in various places, etc.). The imbricate maidenhair fern (*Adiantum pedatum* ssp. *calderi* Cody is common in seeps on serpentine in northwestern California and northward; ssp. *aleuticum* (Rupr.) Calder and Taylor is rather widespread elsewhere in California on nonserpentine substrates. Cody (1983), in assigning the imbricate, congested forms of *A. pedatum* to ssp. *calderi*, notes that it is widespread on serpentines of northeastern North America. Of the two specimens from the west that he cites, one is from serpentine in Del Norte County.

Two other forbs, *Darlingtonia californica* Torr. (the pitcher plant) and *Xerophyllum tenax* (Pursh) Nutt. (bear grass), both so common on serpentine in northwestern California, are further examples of taxa that are good indicators of serpentine habitats, yet are rather common elsewhere on other substrates, albeit often rather sterile (*Darlingtonia* in dune-slack bogs, and *Xerophyllum* on the acidic sands of the Mendocino pine barrens).

There is a second type of distribution pattern where the taxon on serpentine merits indicator status. Several taxa that are widespread on both nonserpentine and serpentine soils can show local or regional prominence on serpentine. Here it is not a matter of exceptional range extensions onto serpentine, but rather that these plants, recruited from nearby sources, come to assume a dominant or indicator status on serpentine. This floristic phenomenon (the "local indicators" of Peterson, 1971) is most often manifested in regions where forest vegetation holds sway. When forest gives way on serpentine to other vegetation types, such as chaparral woodland or grassland, part of the floristic assemblage on serpentine will be made up of those common species that act as local indicators of serpentine. Examples of this category of floristic affinity can be found among each of the major life-forms.

Two conifers typify this indicator role. Digger pine (*Pinus sabiniana*) occurs widely throughout cismontane central and northern California, predominantly in the foothills bordering the Central Valley. Where serpentine outcrops in the midst of yellow-pine or mixed-evergreen forest, scattered individuals of digger pine stand out as clear markers of the substrate. Places where this can be seen are abundant on the serpentines of northern Napa, Lake, Colusa, and Mendocino counties and in the Sierra Nevada. Serpentine appears to extend upward the altitudinal limits of digger pine.

Another pine, *Pinus attenuata* (knobcone pine), is noted for its occurrences on serpentine as just one of its preferences for a variety of azonal soils. Knobcone pine occurs in many California plant communities, from southern California to northwestern California and southwestern Oregon; its populations are usually on disjunct islands of some stressful substrate. After its southernmost serpentine occurrence in the Santa Ana Mountains (Vogl 1973), it seems not to reappear on serpentine until it reaches the North Bay counties. It comes into its own as a serpentine indicator in northwestern California and southwestern Oregon, but it is discontinuously encountered on nonserpentine substrates throughout its range.

Certain wide-ranging shrub species can become indicators of serpentine throughout part or all of their ranges. Where serpentine evokes a local chaparral island in a forest climate, one can expect common chaparral shrubs to appear: *Ceanothus cuneatus, Adenostoma fasciculatum*, some species of *Arctostaphylos, Heteromeles arbutifolia*, and a shrubby variant of *Umbellularia californica* (H. and A.) Nutt. are examples. These taxa are often co-dominant with the serpentine-endemic shrubs discussed earlier. Attempts to demonstrate that the nonendemic shrubs on serpentine are ecotypically (genotypically) adapted variants have so far been inconclusive.

INDIFFERENT OR BODENVAG (UBIQUIST) SPECIES

The third broad mode of occurrence on serpentine is not sharply set off from the grouping just discussed. Some taxa in the California flora occur on and off serpentine repeatedly throughout their ranges. In any given locality, such taxa are found on adjacent nonserpentine substrates as well as on serpentine. Students of edaphic specialization have categorized such plants as *indifferent* (Krause, 1958), *ubiquist* (Krause, 1958), or *bodenvag* (Unger, 1836; Kruckeberg, 1969b). The terms are essentially synonymous and are meant to apply to taxonomic indifference: the same taxon occurs on contrasting substrates, the serpentine occurrences not meriting taxonomic recognition. Not to be overlooked is the possibility that such indifferent taxa may well be ecotypically differentiated into serpentine-tolerant and -intolerant races (e.g., Kruckeberg, 1954). The bodenvag phenomenon manifests itself in many parts of the world where special substrates like serpentine occur.

Bodenvag taxa should be further distinguished from the other two categories of taxa occurring on and off serpentine. The bodenvag taxon appears on both sides of an edaphic contact, seemingly indifferent to the substrate differences. Unlike the first two categories, members of the bodenvag group do not appear on serpentine as altitudinal or geographic disjuncts, nor do they attain indicator status by their relative dominance on serpentine. In a nutshell, they are indifferent to substrate contrasts in both a taxonomic and phytogeographic sense.

The frequency of occurrence of this phenomenon is strongly associated with the severity of the serpentine environment: The more stressful —and barren—the serpentine site, the less common are the bodenvag species. Insolation, soil moisture, soil depth and stabilty, slope exposure, and nutrient status are integrated at various levels to create degrees of stress. Moreover, serpentine outcrops shift from xeric to mesic conditions as one proceeds from south to north. A serpentine barren in the redwood belt will be under less stress than one in the drier inner south Coast Ranges. Hence North Coast serpentine localities will have more bodenvag plants than will, say, New Idria in the south Coast Range.

Each of the primary life-forms in the California flora— tree, shrub, herb—has species or infraspecific variants with this character. I have compiled a list (Table 4) of some possible (to probable) bodenvag species, with samples selected from the major life-forms and plant families; taxa marked with an asterisk have been confirmed as occurring on serpentine. All the others are likely to have serpentine populations, but are not yet confirmed.

Members of the family Compositae are particularly well represented on this list, as are most of the larger grass genera. A number of introduced annual grasses must be included as frequent bodenvag species: *Bromus mollis*, *B. rubens*, *Festuca megalura* Nutt., *Avena fatua*, *Lamarckia aurea* (L.) Moench., *Aira* spp., *Phleum pratense* L., *Gastridium ventricosum* (Gouan) Schinz & Thell., *Polypogon* species, and likely other annual grasses. Occasionally, introduced dicot herbs appear on serpentine; representative examples include several composites (*Hypochaeris radicata* L., *Centaurea* spp., *Lactuca serriola* L.) and other widespread weeds (e.g., *Rumex acetosella*, *Cerastium* spp., *Sagina*, *Spergularia*, *Brassica*, *Trifolium*, *Vicia* and *Erodium*). It should be remarked that ruderals and other introduced weeds are mainly restricted to disturbed serpentine sites (especially roadcuts and banks); only the weedy bromes seem to have been able to succeed on relatively undisturbed bare ground, growing side by side with serpentine indicators or endemics amongst shrubs of serpentine chaparral type.

Wide-ranging species that live in a variety of environments may be expected to have either a broad phenotypic/genotypic tolerance (the ''general-purpose'' genotype of Baker, 1965), or to have evolved a local and differential genotypic accommodation to each variant habitat. The latter response, ecotypic variation, has been demonstrated for several bodenvag species (Kruckeberg 1951, 1954, 1967; Main 1974). Serpentine-tolerant populations are known for *Gilia capitata* Sims., *Salvia columbariae* Benth., *Achillea lanulosa* Nutt., and some other herbaceous dicots (Kruckeberg 1951, 1954). The one grass tested (*Sitanion jubatum* J.G.

Table 4. Examples of taxa indifferent to serpentine (bodenvag taxa)[a]

Ferns

  Pityrogramma triangularis
  *Pteridium aquilinum
  Polystichum imbricans

Conifers

  Abies concolor
  *Abies magnifica shastensis
  *Pinus monticola
  *Pinus contorta
  *Pinus albicaulis
  *Pseudotsuga menziesii
  *Chamaecyparis lawsoniana

Monocots

  *Chlorogalum pomeridianum
  *Fritillaria spp.
  Zygadenus fremontii
  Goodyera oblongifolia
  *Brodiaea spp.
  *Allium spp.
  *Iris spp.
  *Yucca whippleyi var. *percursa*

Grasses and graminoids

  Stipa spp.
  Calamagrostis spp.
  Agrostis spp.
  Danthonia spp.
  Poa spp.
  Melica spp.
  Festuca (annual and perennial spp.)
  Bromus spp.
  *Sitanion jubatum
  Hordeum spp.
  *Elymus glaucus
  Agropyron spp.
  Carex spp.
  Juncus spp.

Woody dicots

  Arbutus menziesii
  *Umbellularia californica
  *Arctostaphylos spp. (e.g., *A. glandulosa, A. viscida*)
  Ledum glandulosum var. *columbianum*
  *Leucothoe davisiae
  *Rhododendron occidentale

  *Eriodictyon californicum
  Chrysothamnus nauseosus ssp. *albicaulis*
  Chrysothamnus parryi
  Salix spp.
  Chrysolepis sempervirens
  Quercus sadleriana
  Quercus garryana breweri
  *Dendromecon rigida
  Ribes spp.
  Amelanchier alnifolia
  Cercocarpus betuloides
  Holodiscus discolor
  Rosa spp.
  Sorbus spp.
  *Cercis occidentalis
  *Pickeringia montana
  *Ceanothus spp.
  *Rhamnus spp.
  Rhus diversiloba
  Garrya spp.
  *Umbellularia californica (shrubby form)
  *Calycanthus occidentalis

Herbaceous dicots

  *Dodecatheon clevelandii var. *patulum*
  Convolvulus malacophyllus
  Asclepias cordifolia
  *Gilia capitata
  Linanthus spp.
  *Phacelia (annual spp.)
  *Phacelia (perennial spp.—e.g., *P. heterophylla, P. californica, P. imbricata*)
  Cryptantha spp.
  *Salvia columbariae
  Salvia sonomensis
  Scutellaria spp.
  *Stachys rigida var. *quercetorum*
  Trichostemma spp.
  *Castilleja spp. (e.g., *C. foliosa*)
  Collinsia spp.
  Cordylanthus spp. (e.g., *C. pilosus, C. tenuis, C. bolanderi*)
  Mimulus (annual spp.)
  Mimulus (perennial spp.)
  Eriogonum spp.
  *Polygonum spp.

Table 4. (cont.)

Claytonia lanceolata
*Silene spp. (e.g., *S. californica,*
   S. multinervia)
Cerastium arvense
Arenaria spp.
Thlaspi montanum
*Streptanthus tortuosus
*Erysimum spp.
*Arabis spp.
*Eschscholtzia spp.
*Platystemon californicum
Sedum spp.
Fragaria spp.
Horkelia spp.
Potentilla spp.
Astragalus spp.
Lotus spp.
*Lupinus spp.
*Trifolium spp.
*Viola spp.
Polygala spp.
Limnanthes spp.
Navarretia spp.
Sidalcea spp.
*Camissonia spp.
*Clarkia spp.
Epilobium spp.
Gayophytum spp.
*Daucus pusillus
*Lomatium spp.
Perideridia spp.
*Pinguicula vulgaris
*Plantago erecta (=P. hookeriana
  var. *californica)*
Githopsis spp.
Downingia spp.
*Achillea lanulosa
Solidago spp. (e.g., *S.*
  multiradiata)
Lessingia spp.
*Haplopappus spp. (e.g., *H.*
  arborescens)
Erigeron spp.
Chrysopsis spp.

Aster spp.
*Crepis spp.
Microseris spp.
Malacothrix spp.
Hieracium spp.
Agoseris spp.
Stephanomeria spp.
Cirsium spp.
Brickellia spp.
Helenium spp.
*Eriophyllum spp.
*Lasthenia californica
*Chaenactis spp.
Blennosperma nanum
*Rudbeckia californica
   var. *intermedia*
Wyethia spp.
*Helianthus gracilentus
*Helianthus bolanderi
Balsamorhiza spp.
Gnaphalium spp.
Antennaria spp.
Layia spp.
*Calycadenia spp.
Madia spp.
Senecio spp.
*Arnica spp.

[a]Genera listed with species (e.g., *Stipa* spp.) indicates that more than
one species in the genus may have <u>bodenvag</u> status.

*Confirmed occurrences by the author or other informant.

Sm.—Kruckeberg, 1951) showed no difference in tolerance, suggesting that nonserpentine populations are preadapted for serpentine. The only other plant I know of that is uniformly tolerant to extremes of edaphic variation is *Typha latifolia* L. (McNaughton et al., 1974).

Bodenvag grass species can be shown to have serpentine-tolerant and -intolerant races; Main (1974) determined by experiment that *Agropyron spicatum* (Pursh) Scrib. and Sm. has developed ecotypic variants for living on or off serpentine. Woody species, both trees and shrubs, should be expected to reveal similar ecotypic variation, but the only positively known case of racial differentiation of a wide-ranging woody plant is *Pinus contorta* Dougl. ex Loud. (Kruckeberg, 1967). In contrast, Griffin (1965) was not able to demonstrate ecotypic variation among serpentine and nonserpentine populations of *Pinus sabiniana*.

EXCLUSION: AVOIDANCE OF SERPENTINE

Contact zones between serpentine and nonserpentine soil types are often sharply discontinuous in both vegetation cover and life-form as well as in floristic composition. The latter attribute merits closer inspection in this section. Many species present on the nonserpentine side of a contact fail to appear on the serpentine. Two examples of this phenomenon may suffice to emphasize the marked change in species composition that is found on either side of a contact, not only in California but in most serpentine habitats elsewhere.

The oak-woodland formation, so common along the Sierra foothills and at low elevations in the Coast Ranges is characterized by several dominant woody species and a rich grass–forb cover. Serpentine intrudes throughout the oak woodland from San Benito County to northwestern California, but the dominant hardwoods fail to cross the contact zone onto serpentine: *Quercus douglasii* H. and A., *Q. agrifolia* Nee., *Q. wizlizenii* A. DC. (to a lesser extent), *Arbutus menziesii* Pursh, and *Aesculus californica* (Spach) Nutt. clearly display this avoidance reaction. The only tree species that commonly grows on both sides of the contact is *Pinus sabiniana*. Also largely absent on serpentine are most of the herbaceous species, especially the native forbs which form a rich ground layer in the oak woodland on nonserpentine soils. *Avena fatua* is usually excluded from serpentine except where the habitat is disturbed (roadcuts, borrow pits, etc.).

The pronounced exclusion of oak-woodland species from serpentine has another consequence: the serpentine flora with its xeromorphic shrubs and sparse groundcover effectively shifts the vegetation type to a serpentine chaparral formation sharply bounded by oak woodland nearby.

Less often, edaphic contacts support contrasting chaparral communities. Differences in species composition on either side of the contact can be substantial. Moving from nonserpentine to serpentine chaparral, species replacement within genera may occur, or genera represented on nonserpentine may be absent on serpentine. In the former category are replacements within *Quercus* (*Q. dumosa* and *Q. wizlizenii* by *Q. durata*), *Ceanothus* (*C. purpureus* Jeps. by *C. jepsonii*), *Arctostaphylos* (*A. canescens* and *A. glandulosa* Eastw. by *A. montana*), and *Garrya* (*G. fremontii* Torr. by *G. congdonii*). Instances of the exclusion of genera in the adjacent nonserpentine chaparral include *Prunus, Heteromeles, Cercocarpus*, and *Pickeringia*.

Being excluded from serpentine may be more than a matter of certain excluded species within bodenvag genera. In the larger sense of contrasting serpentine and nonserpentine floras of the California Floristic Province, we can single out whole genera that are singularly lacking in serpentine tolerance. The genera listed in Table 5 have taxa on nearby nonserpentine areas or have climatic tolerances congruent with those of serpentine plants. Their notable paucity or absence on serpentines suggests that such genera may not have the necessary genotypic preadaptedness to gain a foothold on serpentine.

In larger genera, avoidance need not be absolute. Some few taxa in such genera may show varying degrees of fidelity to serpentine, ranging from one or two endemics to a few bodenvag species. However, most of the constituent taxa are excluded from serpentine. Examples of such genera include *Delphinium, Ranunculus, Arabis, Phlox, Gilia, Cryptantha, Castilleja, Orthocarpus, Trifolium, Lupinus, Astragalus, Lotus,* and several genera of the tribe Cichorieae (Compositae).

Avoidance, as a response to stressful edaphic situations, has been commented on by others. Reference to calciphobes is common in the literature on limestone vegetation and avoidance is observed as a major response in habitats where heavy metal toxicity occurs. Antonovics et al. (1971) have noted that some species adjacent to mine tailings saturated with heavy metals cannot genetically adapt to the stressful sites. Only a small proportion of herbaceous perennials acquire ecotypic tolerance.

Why some genera or species and not others can muster the genetic resources to adapt to serpentine, etc., is a mystery. I am tempted to offer a hypothesis to explain these non-occurrences: In order for a nonserpentine species to acquire serpentine tolerance, directional selection must occur to move a segment of the nonserpentine species in the direction of enhanced serpentine tolerance. For this to occur, the necessary genetic variation—of the appropriate physiological attributes—must be in the gene pool. Neo-Darwinian theory supposes that any such preadaptedness might occur fortuitously before selection for serpentine tolerance occurs. Only if such latent variation is available will selection for tolerance occur. I would contend that it is a matter of chance as to the availability or occurrence of the requisite genetic capacity.

CALIFORNIA SERPENTINE FLORA BY PHYSIOGRAPHIC PROVINCE

The floristic response to serpentine varies from place to place in California. First, there is the shift in floristic composition, life-form, and physiognomy of the vegetation from south (Santa Barbara and San Luis Obispo counties) to north (northwestern California at the Oregon border). This trend is at first revealed as a shift toward a more mesic-appearing plant cover. As the serpentine-barren phenomenon decreases in severity, a woody, often arboreal, plant cover increases. A similar trend obtains between coastal and inland outcrops. This trend is most pronounced north of San Francisco Bay. Serpentine outcrops in the coastal redwood belt appear to have more ample plant cover (and biomass) than those further inland, as along the dry western flank of the Great Valley.

There is an evident increase in the number of endemics and indicator species from south to north. The areas of richest serpentine endemism are in northwestern California (especially in the Klamath–Salmon–Eel river drainages).

These changing patterns of floristics and vegetation on serpentine can be seen through by a closer analysis of the more salient botanical features of each major physiographic province in the state.

Table 5. Some genera that largely avoid serpentines in California

Ferns

Asplenium
Athyrium
Blechnum
Cheilanthes (s.s.)
Cryptogramma
Cystopteris
Dryopteris
Pellaea
Woodsia

Conifers

Picea
Thuja
Torreya
Tsuga
Sequoia
Sequoiadendron

Woody dicots

Acer
Aesculus
Alnus
Amelanchier
Amorpha
Baccharis
Cassiope
Cercis
Cercocarpus
Chamaebatia
Chrysothamnus
Clematis
Cornus
Corylus
Dirca
Euonymus
Fraxinus
Gaultheria
Helianthemum
Holodiscus
Lonicera
Menziesia
Osmaronia
Paxistima
Philadelphus
Phyllodoce
Physocarpus
Pickeringia
Populus
Ptelea

Ribes
Rosa
Rubus
Sambucus
Sorbus
Spiraea
Staphylea
Vaccinium
Whipplea

Dicot herbs

Achlys
Agoseris
Angelica
(*Amaranthaceae* genera)
Amsinckia
Apocynum
Aralia
Aristolochia
Artemisia
Aruncus
Asarum
Boisduvalia
Brickellia
Cacaliopsis
(all *Cactaceae*)
Calandrinia
Calyptridium
Campanula
Cardamine
Caulanthus
(all *Chenopodiaceae*)
Coptis
Crepis?
Crocidium
Cynoglossum
Descurainia
Draba
Draperia
Eryngium
Eupatorium
Euphorbia
Filago
Fragaria
Gayophytum
(all *Gentianaceae*)
Geranium
Geum
Gilia (except
  G. capitata)

Gnaphalium
Grindelia
Hackelia
Heracleum
Hesperochiron
Hydrophyllum
Hypericum
Ipomopsis
Ivesia
Kelloggia
Lathyrus
Lepidium
Ligusticum
Linnaea
Lithophragma?
Lithospermum
Lotus
Luetkea
Luina
Malacothrix
Marah
Mentzelia
Mertensia
Micropus
Microsteris?
Mirabilis
Nemacladus
Nemophila
Oenothera (s.s.)
Orthocarpus?
Osmorhiza
Oxalis
Paeonia
Pedicularis
Penstemon
Petasites
Phlox
Plectritis
Polemonium
Potentilla
(all *Primulaceae?*)
Pteryxia
(all *Pyrolaceae?*)
Ranunculus
Rumex
Satureja
Saussurea
(all herbaceous
  Saxifragaceae)
Thalictrum
Thelypodium?

Table 5. (cont.)

Thermopsis	*Juncus*
Thysanocarpus	*Koeleria*
Trautvetteria	*Listera*
Trifolium	*Luzula*
Valeriana	*Lysichiton*
Vicia	*Maianthemum*
Zauschneria	*Muhlenbergia*
	Narthecium
	Orcuttia?
Monocot herbs	*Panicum?*
	Phleum
Agrostis	*Pleuropogon*
Aristida	*Puccinellia*
Brodiaea	*Schoenolirion*
Camassia	*Scirpus*
Carex (largely unknown,	*Scoliopus*
mostly avoiders)	*Sisyrinchium?*
Clintonia	*Smilacina*
Cyperus	*Stenanthium*
Danthonia?	*Stipa?*
Deschampsia	*Streptopus*
Disporum	*Tofieldia*
Glyceria	*Trillium*
Habenaria?	*Trisetum*
Heleocharis	*Veratrum*
Iris	

South Coast Ranges

This southern mountain complex is considered here to extend from Santa Cruz and western Stanislaus counties south to the San Rafael Mountains of Santa Barbara County. This vast and geologically complex network of partially isolated mountain ranges contains abundant serpentine outcrops, both in more moist coastal regions and in the harsh, dry interior mountains like the Temblor and Diablo ranges.

I estimate that there are 36 taxa endemic to serpentine in the south Coast Ranges. Examples include the bizarre tarweed *Layia discoidea* Keck of the New Idria barrens and the unique crucifer *Streptanthus insignis* Jeps.,with its showy terminal "flag" of purplish black sterile flowers, as well as *Allium howellii* Eastw. var. *sanbenitensis* (Traub) Ownbey and Aasc, *Arctostaphylos obispoensis* Eastw., and *Carex obispoensis* Stacey. *Cupressus sargentii*, a serpentine endemic of wider Coast Range distribution, reaches its southern limit in this region, in the San Rafael Mountains of Santa Barbara and San Luis Obispo counties. (Appendix E lists the known and probable endemics of the south Coast Range serpentines.)

It has been remarked by others (e.g., J.R. Griffin and V.L. Holland, personal communications) that some taxa of wider distribution on a variety of substrates only occur in the south Coast Ranges on serpentine. Examples of such restriction include *Agropyron trachycaulum* (Link) Malte., *Melica stricta* Bol., *Calochortus invenustus* Greene, *Salix coulteri* Anderss., *Quercus dumosa* (mainly on serpentine in San Luis Obispo County,—Holland, personal communication), *Lewisia rediviva* Pursh, *Dudleya blochmanii* (Eastw.) Moran, *Fremontia californica* Torr. subsp. *obispoensis* (Eastw.) Munz, *Linum perenne* L. subsp. *lewisii* (Pursh) Hult., *Viola purpurea* Kell. subsp. *mohavensis* J. Clausen, *Perideridia pringlei*

(Coult. and Rose) Nels. and Macbr. (L. Constance), *Acanthomintha obovata* Jeps., and *Salvia sonomensis* Greene (Holland).

San Francisco Bay Region

Probably the first recognition of serpentine in California can be claimed for the Bay Area, through the cartography of early explorers (Norris and Webb, 1976, p. 572). Bay Area botanists of the 19th century should have noticed the distinctive floras on the many nearby outcrops. But such recognition, at least in print, had to wait until the mid-1940s. Although many serpentine exposures in proximity to population centers have disappeared under buildings and roads, samples of urban serpentine do still exist. Witness the fine exposures at the Presidio in San Francisco, on Tiburon peninsula, and in the Oakland and Berkeley hills.

For us, the Bay Region here includes the counties bordering San Francisco Bay (Marin, Contra Costa, Alameda, San Francisco, and San Mateo counties); for Santa Clara County, see the previous section on the south Coast Ranges.

Several areas of endemism occur within the Bay Region (see Appendix E). Three of the most striking and well-known are: the Presidio within San Francisco (e.g., *Arctostaphylos hookeri* G. Don subsp. *franciscana* (Eastw.) Munz and *Clarkia franciscana* Lewis and Raven), the Tiburon peninsula (e.g., *Castilleja neglecta, Streptanthus niger, Calochortus tiburonensis*), and Mt. Tamalpais, dominating the Bay Area skyline above its north shores (e.g., *Streptanthus batrachopus, S. glandulosus* subsp. *pulchellus* (Greene) Kruckeberg, and *Cirsium vaseyi* (Gray) Jeps.). The serpentines of the Crystal Springs area in San Mateo County have been spared destruction largely through the protection offered by their inclusion in the domestic water-reservoir lands which are off limits to the public. A few endemics are found here: *Cirsium fontinale* (Greene) Jeps., *Eriophyllum latilobum* Rydb., and *Acanthomintha obovata* Jeps. subsp. *duttonii* Abrams.

Linum (Hesperolinon) congestum Gray is endemic to several Bay Area serpentine outcrops. In all, I calculate that there are 19 taxa endemic to serpentine in the Bay Area. They are either in serpentine grassland communities or in serpentine chaparral/chaparral woodland. Common associated species for grassland serpentine are: *Stipa* spp., *Calamagrostis ophitidis* (J.T. Howell) Nygren (restricted to serpentines of Marin and Lake counties), *Zygadenus fontanus* Eastw., *Brodiaea peduncularis* (Lindl.) Wats. (?), and *Eriogonum vimineum* Dougl. ex Benth. var. *caninum* Greene (Howell, 1970, p. 17). In chaparral woodland on serpentine, *Cupressus sargentii, Quercus durata, Ceanothus jepsonii*, and *Arctostaphylos montana* are frequent woody members of the community; *A. montana* is restricted to Marin County.

I am not aware of any unusual range extensions or plants of indicator status for Bay Region serpentines.

North Coast Ranges

The magnitude of serpentine outcrops increases significantly as one proceeds northward from Napa and Sonoma counties all the way to the Oregon border. Indeed, serpentine lithology dominates much of the terrain of the northwestern counties of California, especially the northernmost ones (Del Norte, Trinity, western Siskiyou, and western Shasta counties). I reckon that there are 90 to 100 taxa endemic to serpentine and peridotite within this vast mountainous, predominantly forested, region (see Appendix E).

Though somewhat artificial, it seems convenient to subdivide the north Coast Ranges into three sectors; some endemic taxa occur in two or all three of the subdivisions, despite the partial physiographic discontinuity.

a. The Napa–Sonoma–Lake counties area

This region corresponds approximately with the Santa Rosa sheet of the Geological Map of California (Koenig, 1963). In Sonoma County the major serpentine areas are in the northwestern and northeastern sections of the county. A massive outcrop called "The Cedars" is in the upper Austin Creek drainage north of Cazadero. A cedar woodland (pure stands of *Cupressus sargentii*) harbors endemics nearby; the most notable is *Streptanthus morrisonii* Hoffman. The northeast section of the county, in the vicinity of The Geysers in the Mayacamas Mountains, is the endemic home of *Streptanthus brachiatus*. Here the serpentine outcrops extend from Mt. St. Helena in northwest-trending bands.

The Napa County serpentines are mainly north and east of Mt. St. Helena; the country from St. Helena Creek and the crest of the Mayacamas Mountains east to Lake Berryessa is rich with serpentine exposures. *Streptanthus hesperidis, S. barbiger* Greene, *S. brachiatus*, and *S. morrisonii* Hoffman var. *elatus* Hoffman richly represent this cruciferous genus in this region. The larger Napa County outcrops continue northwestward into southeastern Lake County, supporting several of the same endemics.

I estimate there to be about 27 taxa endemic to serpentine in the Napa–Sonoma–Lake counties region. Besides the *Streptanthus* species cited above, other notable endemics include *Erythronium helenae* Appleg., *Senecio clevelandii* Greene, *Nemacladus montanus* Greene, *Madia hallii* Keck, *Cordylanthus pringlei* Gray, *Mimulus nudatus* Curran ex Greene, *Monardella villosa* Benth. ssp. *neglecta* (Greene) Epl., *Cryptantha hispidula* Greene ex Brand, *Arctostaphylos stanfordiana* Parry ssp. *bakeri* (Eastw) Adams, *Lomatium ciliolatum* Jeps. var. *hooveri* Math. and Const., *Ceanothus jepsonii* var. *albiflorus* J.T. Howell, and *Delphinium uliginosum*. Serpentine outcrops in these three counties are usually at moderate elevations (1000–3000 ft.) and adjacent to (associated with) several vegetation types (Kuchler, 1977) on normal soils. To the west, redwood forest and mixed hardwood forest can surround serpentine and its hard chaparral or cypress–scrub stands. In the drier interior, from the Mayacamas Mountains to Lake Berryessa, the dominant vegetation on nonserpentine soils is either chaparral (*Adenostoma/Arctostaphylos/Ceanothus*) or blue oak–digger pine woodland. These nonserpentine vegetation types give way to serpentine chaparral and cypress groves on ultramafics, much like the western Sonoma County serpentines. Often where the blue oak–digger pine type prevails on nonserpentine, the digger pine persists as scattered individuals within the serpentine chaparral. Vernal pools, seeps, and streams are frequent on serpentine areas throughout the region. These moist habitats can support a unique mix of plants, some endemic (*Senecio clevelandii, Mimulus nudatus*), others of nonserpentine affinity but thriving in wet serpentine places (*Rhododendron occidentale* (T. and G.) Gray, *Calocedrus decurrens, Calycanthus occidentalis* H. and A., and *Aquilegia eximia* Van Houtte ex Planch).

b. Mendocino, Colusa, Tehama, and Humboldt counties

This extensive region lies roughly in the middle of the north Coast Ranges, between the Klamath Ranges to the north and the Sonoma–Napa–Lake counties area to the south. The

Ukiah (Jennings and Strand, 1960) and Redding (Strand, 1962) sheets of the Geologic Map of California depict massive outcrops of serpentine for these counties. The major outcrops are confined to two rather distinct climatic zones; those above the west flank of the Great Valley from Highway 20 (at the Lake–Colusa counties line) north to Paskenta are driest and are dominated by serpentine chaparral. Westward into the more moist and higher elevations of the north Coast Ranges (e.g., Snow Mountain, the Yolla Bolly Mountains), the serpentine supports chaparral woodland or open Jeffrey pine–incense cedar forest in the midst of yellow pine–fir forests on nonserpentine.

Known endemics to serpentine currently total about 15 taxa, though the efforts of botanists at Humboldt State University will surely see that number increased. Outstanding examples include *Haplopappus ophitidis* (J.T. Howell) Keck, *Asclepias solanoana* Woodson, *Calochortus coeruleus* (Kell.) Wats., *Allium hoffmanii* Ownbey, and *Streptanthus drepanoides* Kruckeberg and Morrison.

c. The Klamath Ranges (Del Norte, Trinity, Siskiyou, and Shasta counties of California, and Jackson, Josephine, and Curry counties of Oregon)

The California Floristic Provinces of Raven and Axelrod (1978) extend into southwestern Oregon to accommodate the rich floras of the Klamath Ranges. Thus this center of endemism in California is shared with Oregon, with a high number of relict taxa. Great stretches of serpentine and other ultramafics intrude other lithologies in this mountainous region of complex geology. For northwestern California alone, serpentine endemics number 19 taxa; when we include the contiguous counties of southwestern Oregon, the number increases to over 30 taxa.

Some of the more striking serpentine endemics include: *Lilium bolanderi* Wats., *L. kelloggii* Purdy, *Allium hoffmanii, Calochortus caeruleus, Fritillaria glauca* Greene, *Erythronium citrinum* Wats., *Streptanthus barbatus* Wats., *S. howellii* Wats., *Arabis macdonaldiana* Eastw., *A. aculeolata* Greene, *Lomatium engelmannii* Math., *Perideridia leptocarpa* Chuang and Const; *Asclepias solanoana, Phacelia dalesiana* J.T. Howell, *Haplopappus ophitidis*, and *Arnica cernua* Howell. (A full listing of the endemics of this remarkable region is in Appendix E.)

Sierra Nevada

The overall effect of serpentine on floras in the Sierra Nevada parallels that of the inner Coast Ranges. Where forest reigns on nearby nonserpentine sites, the serpentine supports chaparral with scattered conifers (chaparral woodland). Blue oak–digger pine communities on nonserpentine also give way to chaparral or grassland on serpentine. And nonserpentine chaparral or grassland is replaced by a sparse rocky grass–forb community of low productivity on serpentine.

Though there are superb samples of sere, barren serpentines in the Sierra, from Tulare County in the south to the Feather River country in the north, the number of endemics (16 taxa) is much lower than that (150 taxa) for the Coast Ranges. Exposed serpentines are no younger (late Pliocene—Raven and Axelrod, 1978) than most serpentines of the Coast Range. Other differences may contribute to understanding the disparity in the number of endemics between the two regions:

1. Sierra Nevada serpentines are restricted mainly to the lower western slopes of the

range, and thus are predominantly within climatic zones that support similar vegetation: blue oak and digger pine or chapparral, and grassland. It is only in the northern sector that serpentine occurs in altitudinally higher forested areas. In contrast, Coast Range serpentines are subjected to markedly different climates, even at a given latitude; this is manifested in the differences between the outer and inner Coast Ranges. There is nothing in the Sierra like these great differences in topographic exposure of northwestern California, where wet shifts to dry in rain-shadow areas.

2. Sierra Nevada serpentines are somewhat more contiguous throughout their narrow northwest-trending belt. Such contiguity of serpentine substrate might have lessened the opportunity for local, narrow endemism.

Outstanding examples of taxa endemic to Sierra Nevada serpentines include *Chlorogalum grandiflorum* Hoov., *Allium sanbornii* Wood, *Streptanthus polygaloides* (perhaps the most distinctive member of the genus), *S. tortuosus* Kell. var. *optatus* Jeps; *Lupinus spectabilis* Hoov., *Lomatium marginatum* (Benth.) Coult. and Rose var. *marginatum, Calystegia stebbinsii* Brommitt, *Cryptantha mariposae* Jtn., and *Senecio lewis-rosei* J.T. Howell.

A limited number of taxa appear as local or regional indicators of serpentine, in addition to obligate serpentine endemics. Examples of such taxa are *Parvisedum pumilum* (Benth.) R.T. Clausen, *Clarkia arcuata* (Kell.) Nels. and Macbr., *Orthocarpus lacerus* Benth., and *Calycadenia truncata* D.C. var. *scabrella* (E. Drew) Keck. It is possible that the rare Sierran shrub *Fremontodendron decumbens* Lloyd also fits the indicator category. Since its first discovery at Pine Hill, El Dorado County (Lloyd, 1965), where it is restricted to an olivine gabbro (mafic) substrate, it has since been reported from nearby serpentines.

Floristic affinities between the Sierra Nevada and the Coast Ranges are not uncommon. The contemporary connection is, of course, in northern California, where suitable forest and chaparral bordering the Great Valley extend between the two mountain systems. Yet it must be remembered that serpentine outcrops are by no means continuous across this northern "bridge."

For serpentinicolous species, the Coast Range– Sierra Nevada connection is substantial. Some wide-ranging serpentine endemics are shared by the two ranges. Woody plants restricted to serpentine in both the Sierra Nevada and Coast Ranges include: *Cupressus macnabiana, Quercus durata, Lithocarpus densiflorus* (H. and A.) Rehd. var. *echinoides* (R. Br.) Abrams, and *Garrya congdonii*. Examples of restricted herbaceous plants are: *Chlorogalum angustifolium* Kell., *Balsamorhiza macrolepis* Sharp, *Cirsium breweri* (Gray) Jeps., *Helianthus exilis* Gray, *Lagophylla minor* (Keck) Keck, *Monardella villosa* Benth. ssp. *sheltonii* (Torr.) Epl., *Dentaria pachystigma* Wats. var. *dissectifolia* Octl., *Eriogonum tripodum* Greene, and *Polygonum spergulariaeforme* Meissn.

Taxa that serve as local indicators of serpentine may occur in both ranges. Some are nonserpentine in the Sierras but are faithful to serpentine in the Coast Ranges (e.g., *Pinus jeffreyi, P. balfouriana* Grev. and Balf., *Calocedrus decurrens, Eriogonum pyrolaefolium* Hook., *Ivesia gordonii* (Hook.) T and G., and *Quercus vaccinifolia*). Others may "indicate" serpentine in both mountain systems, while still retaining nonserpentine populations in both areas. Examples of this type include *Aspidotis densa, Polystichum scopulinum* (D.C. Eat.) Maxon, *Umbellularia californica* (H. and H.) Nutt. (shrubby form), *Pinus attenuata* (serpentine in Sierra Nevada?), *Melica stricta, Poa tenerrima* Scribn., *Carex serratodens* W. Boott, *Allium cratericola* Eastw., *Darlingtonia californica, Xerophyllum tenax, Calochortus invenustus, Lewisia leana* (Porter) Rob., *Berberis piperiana* (Abrams) McMinn., *B.*

pumila Greene, *Calycanthus occidentalis, Sedum obtusatum* Gray ssp. *obtusatum, Sidalcea diploscypha* (T. and G.) Gray, *Linum perenne* ssp. *lewisii, Epilobium obcordatum* Gray var. *laxum* (Hausskn.) Dempst., *Gentiana newberryi* Gray, *Dodecatheon clevelandii* Greene ssp. *patulum* (Greene) H.J. Thomps., *Convolvulus malacophyllus* Greene, *Polemonium chartaceum* Mason, *Phacelia divaricata* (Benth.) Gray, *P. egena* (Greene) Const., *P. imbricata* Greene, *Salvia sonomensis, Trichostemma rubrisepalum* Elmer, *Antirrhinum breweri* Gray, *A. cornutum* Benth., *Castilleja foliolosa* H. and A., *C. pruinosa* Fern., *Collinsia sparsiflora* F. and M., *Mimulus douglasii* (Benth.) Gray, and *Coreopsis stillmanii* (Gray) Blake. I am not aware of any taxa that are on serpentine in the Sierra Nevada but are only on nonserpentine substrates in the Coast Ranges.

STATISTICAL SUMMARIES OF PLANT LIFE ON SERPENTINES OF CALIFORNIA

Throughout this section on the floristics of California serpentines, I have referred to estimates of the numbers of taxa in certain categories of serpentine addiction. These estimates, in tabular form (Tables 2–4), have been gleaned from a variety of sources. First, a preliminary search of the two major California floras —Munz and Keck, 1959, and Abrams (1923–1960)—provided entries for a card file of taxa with varying degrees of fidelity to serpentine. In some instances, the authors of these floras directly state the degree of serpentine addiction; in other cases, the serpentine habit can be inferred from habitat or locality information in the diagnosis of the taxa. These first-approximation data were then checked by searching the herbaria at the University of California at Berkeley (UC) and the California Academy of Sciences in San Francisco (CAS). Specialists on particular genera or geographic regions of the state have also been consulted. A number of bits of information have come to me via correspondence from California botanists, amateur and professional, and reports from the California Native Plant Society have been helpful. Further, I have been searching the serpentines of California off and on since 1948, so I have had firsthand contact with many of the taxa reported here.

The quality and accuracy of the numerical and taxonomic summaries can be substantially improved as more botanists report their findings on serpentine floras. As yet no one has published detailed botanical surveys for any serpentine habitats, though they are mentioned in regional floristic treatments by county (e.g., Marin County—Howell, 1970; San Luis Obispo County—Hoover, 1970) or by physiographic province (e.g., Trinity Alps—Ferlatte, 1974; Mt. Diablo—Bowerman, 1944; Mt. Hamilton—Sharsmith, 1945). An invaluable source has been the field notes of Freed W. Hoffman, a rancher and amateur botanist who lived all his life in the proximity of serpentine in northern Sonoma County. Hoffman's passion was searching for members of the cruciferous genus *Streptanthus*, many of which are endemic to serpentine. While ''strep-trekking'' (as he called it), Freed made it a point to collect other interesting plants on serpentine. Besides finding several new *Streptanthus* endemic to serpentine (Hoffman, 1952), he discovered several new serpentine taxa in other genera (e.g., *Allium hoffmanii*). Hoffman's field notebooks, now kept at the herbarium at the University of California, Berkeley, yield a rich source of plant collections made on serpentine; I have utilized these unpublished records in this floristic review.

Summaries have been presented here in three kinds of matrices: 1. floristic status—endemic (universal indicators), local indicators, and indifferent (bodenvag) taxa; 2. serpentine affinity by plant family; 3. relative abundance and serpentine affinity by physiographic-floristic province in California. Besides Tables 2 to 5, more detailed tabulations of serpentine taxa are presented in Appendices C, D, and E.

We end this section on the flora of California serpentines with a resumé of the major impressions gained by the analysis.

1. Restriction to serpentine is manifested both in taxonomic terms and by life-form. Endemic species are frequent; they make up 10% (215 taxa) of the total endemic flora of California (2125 taxa—Raven and Axelrod, 1978). This number is all the more impressive when one considers that serpentine constitutes only 1.0% (2860 out of 285,000 km^2) of the landmass of the state. Chaparral woodland, chaparral, and grassland are the major physiognomic types of serpentine vegetation.

2. The serpentine flora is richer in both endemics and local indicators in the Coast Ranges than in the Sierra Nevada. Two factors that may account for this are the greater homogeneity of the Sierra and its smaller extent of serpentine.

3. The number of serpentine endemics in the Coast Ranges is greater north of the Bay Area, despite the general tendency for the "serpentine effect" to be ameliorated by the more mesic climate. This is probably a reflection of the greater areal extent of serpentine in northwestern California.

4 Serpentine floras in California are composed of endemics (universal indicators), local indicator species, and indifferent taxa. Although the endemics and the local indicator taxa are the most intriguing manifestations, the indifferent species may be more abundant and constitute more of the cover and biomass of serpentine outcrops, especially in mesic areas.

5. Avoidance of serpentine by a large proportion of the regional flora on nearby nonserpentine soils is common to all the major serpentine occurrences in the state. This avoidance can involve entire genera and families (Kruckeberg, 1969b; Proctor and Woodell, 1975).

SERPENTINE FAUNA OF CALIFORNIA

If one views the serpentine habitat as an ecosystem, consideration of the faunal component is essential. Moreover, a serpentine ecosystem with its unique substrate and flora would also be expected to harbor unusual faunistic features. Are there unique, endemic animals restricted to serpentine in California? Are there particular coevolutionary phenomena involving serpentine flora and fauna? Are there unusual trophic relationships between the plants that take up the elements of serpentine substrates and their herbivorous predators? These and similar questions, easily provoked by the unusual features of serpentine ecologies, are largely unanswered as yet.

This dearth of knowledge about plant–animal or animal–animal interactions on serpentine is not limited to California. Proctor and Woodell (1975), in their detailed, worldwide coverage of the ecology of serpentine, remark on the paucity of such literature: "Why the animals of serpentine have been so little studied we do not know, but they offer a fascinating field for further work."

It so happens that the only three papers Proctor and Woodell found to review deal with animals found on California serpentines. Paul Ehrlich and his Stanford group have maintained a continuing effort to learn about the population dynamics of the butterfly *Euphydryas editha*, an insect restricted to (or locally abundant on?) serpentine in the San Francisco Bay Area. One aspect of this study was to determine if the host food-plant *Plantago erecta* Morris., common on the Jasper Ridge serpentines, influences the chemistry of its predator (Johnson et al., 1968). No such association could be established. Ehrlich's group contends that *E. editha* is a serpentine ecotype that is adapted to the low humidity and high temperatures of the serpentine habitat, as well as to the absence of pathogens.

The second paper cited by Proctor and Woodell describes the association of the pocket gopher (*Thomomys bottae*) with a primary food plant, *Brodiaea* spp. (Proctor and Whitten, 1971). The brodiaea corms, particularly abundant on Jasper Ridge serpentine, are the principal food source of the gophers. Though the Jasper Ridge serpentine is fairly high (for a serpentine) in calcium, the gophers must ingest much greater than normal amounts of magnesium per unit of calcium—i.e., 5–10 times more magnesium than calcium. The effect of this unusual diet of high magnesium (and possibly heavy metals?) is not discussed.

The third animal case cited in Proctor and Woodell concerns the population biology of the California salamander, *Ensatina eschscholtzii* (R. Stebbins, 1949). Subspecies of this animal intergrade in the presence of serpentine; how the substrate affects the salamander is not known.

Insects are the most likely faunal group to be influenced by serpentine, either directly through the soil or more probably via serpentine plant life as food resources. A recent study in California (Shapiro, 1981) tells of a remarkable association between pierid butterflies and species of the serpentine endemic *Streptanthus* (Cruciferae). Larvae of *Pieris sisymbrii* and

54

P. sara preferentially feed on serpentine *Streptanthus*. In turn, the plants appear to have evolved a unique, partial deterrent to this predation. Leaves of some species of *Streptanthus* develop opaque, non-green callosities at the tips of the marginal teeth; these apparently mimic eggs and thus deter ovipositing by the butterflies. Shapiro has tested this remarkable case of egg mimicry by a host plant by removing the callosities; oviposition increased as a result.

The only case of heavy metals and magnesium accumulation in herbivores known to me involves mound-building termites on serpentines of the Great Dyke of Rhodesia (Wild, 1975). The workers of two termite species on serpentine contain high levels of nickel and chromium, and the magnesium/calcium ratios in these species are higher than in those on nearby nonserpentine soils. The chemical aspects of plant/animal interactions on California serpentines could prove to be equally intriguing.

There is little doubt that some fascinating relationships between serpentine floras and faunas will be discovered. Among the topics meriting study are: animal life on serpentine in relation to vegetation cover; palatability of serpentine plants; coevolved relationships between animals and plants; effects of the exceptional soil chemistry of serpentines on animal physiology; and others.

THE EVOLUTIONARY ECOLOGY OF SERPENTINE BIOTA IN CALIFORNIA

The moment we turn from description to the analysis–synthesis phase of regional floristics, we are confronted with a central evolutionary question: What are the origins of the diverse elements that constitute the biota adapted to a particular place? Inevitably, this larger question breeds particular questions: How do plants adapt to serpentine? Why are some biota excluded? Are there different modes of origin among the many serpentine endemics? What are the causes of the present distributions of local serpentine indicators, often disjunct from their main populations? Is ecotypic accommodation to serpentine a first step toward speciation?

These questions have been the subject of conjecture and hypothesis for some time now. In the 1940s, G.L. Stebbins (1942) and Mason (1946a,b) considered the problem of the origin and spread of endemics. Since that time, the particular problems attending evolutionary adaptation to serpentine have been taken up by me (Kruckeberg, 1954, 1969b), G.L. Stebbins and Major (1965), Proctor and Woodell (1975), and Raven and Axelrod (1978). I will first attempt to summarize and synthesize the ideas developed by these authors, and then will determine if a variant scheme for the origin of serpentine plants is now required.

No one explanation will suffice to cover the variety of floristic manifestations on serpentine. Serpentine plant life encompasses absolute restriction, either as narrow and local endemism or as regional endemics, through more subtle ecotypic differentiation, to apparent indifference. The endemics may be strikingly unique taxa—*Benitoa occidentalis* (Hall) Keck, *Asclepias solanoana, Hesperolinon* spp., species of the subgenus *Euclisia* (*Streptanthus*)—or merely serpentine vicariants of close nonserpentine congeners. Where restriction is not "absolute," serpentine accommodation can range from taxonomic variants to physiological races.

Such a diverse display of adaptation to serpentines may require different models of evolutionary origin and accommodation. Or, these may represent different stages of adaptation that are following the same evolutionary routes. G.L. Stebbins (1980) reminds us that "Like every other problem of evolution, that of the nature and occurrence of rare species is not a simple one that can be solved by applying indiscriminately one or a few general principles." As a consequence, only a synthetic view of the evolution of any biota, narrow endemics included, can cope with the complexly interacting causes—ecological, genetic, and historical.

Given the diversity of serpentine accommodation, what are the models that can be prescribed to explain them? Some possible schemes include:

1. Biotype depletion
2. "Drift"
3. Catastrophic selection and saltational speciation
4. Standard allopatric speciation with ecogeographic specialization
5. Ecotypic differentiation
6. Hybridization, with or without allopolyploidy

The success of any of these modes of origin must depend upon suitable preadaptation to serpentine. That is, there must be genetic resources in nonserpentine progenitors to provide the initial tolerance to serpentine habitats; plants arriving by any of the several modes of access to such a habitat must have genetic variation to match the challenge of the serpentine environment. This requisite preadaptation probably can be mobilized rapidly, followed by the fixation of new tolerances: the acquisition of heavy-metal tolerance studied by Bradshaw and his coworkers can be experimentally achieved in a single or a few generations (Antonovics et al., 1971). Disruptive selection and ecological isolation may then promote further divergence in other characters. The ingredients for speciation are clearly available, once preadapted genotypes are captured by serpentine.

The preadaptation requirement can be invoked to explain the exclusion of many taxa from serpentine. Lack of preadaptation to the serpentine habitat simply prevents the initial step — incipient serpentine tolerance — that could be followed by other events leading to speciation.

Raven and Axelrod (1978) provide a masterful review of ideas on the evolution of endemics, including those on serpentine; further, they offer a stimulating, original analysis of the various modes of serpentine restriction. They distinguish three types of serpentine restriction that can be interpreted as having different modes of origin.

Group I

"Species of woody plants that occur on ultrabasic rocks and are also widely distributed on other substrates." I have dealt with many of these taxa in this review under the heading of "local indicators" or as bodenvag taxa. Raven and Axelrod's Group I (p. 72, 1978) is exemplified by *Adenostoma fasciculatum*, *Calocedrus decurrens*, *Heteromeles arbutifolia*, *Pinus jeffreyi* and *P. sabiniana*, and the authors speculate on the nature and origin of the serpentine tolerance of these taxa.

There is some reason to believe that woody species may acquire serpentine tolerance much as do the better-known cases of herbaceous perennials like *Achillea lanulosa* Nutt. or annuals like *Gilia capitata* and *Salvia columbariae* (Kruckeberg, 1951, 1954, 1967)—that is, by ecotypic differentiation into serpentine-tolerant races. Evidence for this in woody taxa is equivocal. Griffin (1965) tested serpentine and nonserpentine strains of digger pine (*Pinus sabiniana*) on serpentine substrate and could find no difference in tolerance to serpentine. I tested progeny of several bodenvag shrubs, but could find no ecotypic differences. But similar tests on Pacific Northwest woody bodenvag taxa disclosed ecotypic response in *Pinus contorta*, *Spiraea douglasii* Hook., *Juniperus communis* L. var. *saxatilis* Pall., and *Taxus brevifolia* (Kruckeberg, 1967). It is possible that woody plants can either possess general-purpose genotypes in the sense of Baker (1965), or show the more usual ecotypic response. More progeny testing of Group I species is clearly needed.

Certain taxa on Raven and Axelrod's list of Group I plants (p. 72) merit discussion. *Chamaecyparis lawsoniana* fits well the attributes for Group I; the probable reasons for its serpentine and nonserpentine distribution have been discussed earlier in this review, based on the recent findings of D.G. Zobel and Hawk (1980). The shrubby form of tan-oak, *Litho-*

carpus densiflorus var. *echinoides*, seems a better candidate for Group II (see below). In the North Coast and Klamath ranges, var. *echinoides* is nearly or wholly confined to ultramafics, but in the Sierra it may be less obligate on ultramafics: it is chiefly limited to Tuolumne and Mariposa counties, where serpentine is frequent, and localities in Placer and Butte counties are in the vicinity of serpentine, but unfortunately, ecological data are not given on herbarium specimens for these localities. Listing *Rhamnus californica* Esch. as a group I plant obscures the well-defined variant *R. californica* ssp. *crassifolia* (Jeps.) C.B. Wolf, that appears to be confined to serpentine in the north Coast Ranges.

Group II

"Woody taxa that have a discontinuous distribution on serpentine, and are largely but not wholly confined to it" (Raven and Axelrod, p. 73). Examples include *Cupressus macnabiana* A. Murr., *Quercus durata, Ceanothus jepsonii, C. pumilus* Greene, *Garrya buxifolia* Gray, and *Salix breweri* Bebb. I would consider these as good serpentine endemics, widely and discontinuously distributed, with only an occasional infidelity to serpentine. According to Raven and Axelrod, these Group II plants represent a further intensification of ecotypic accommodation to serpentine, in which the nonserpentine races are largely eliminated. The mode of their origin can then be framed on G.L. Stebbins' (1942) model of progressive biotype depletion. In essence, Group II taxa are relictual, the "paleoendemics" of G.L. Stebbins and Major (1965). Parenthetically, the biotype depletion hypothesis now could be tested by isozyme analysis, as has been done for *Stephanomeria malheurensis* Gottlieb (Gottlieb, 1973). G.L. Stebbins (1980) has modified his biotype depletion hypothesis of 1942, recognizing that depleted gene pools are not confined to rare and local taxa; common inbreeders may also have limited genetic variability. Moreover, G.L. Stebbins (1980, p. 79) reminds us that some rare taxa can have "relatively rich stores of genetic variability." Clearly, the question of origins of rare taxa is a multiple one, requiring a synthetic paradigm.

Group III

Taxa "chiefly herbaceous and wholly confined to ultrabasic rocks, usually in restricted local areas" (Raven and Axelrod, pp. 73–75). These are the narrow edaphic endemics first discussed by Mason (1946a,b) and extensively treated here in the section on "Serpentine Endemism." Besides the good representative sample in Raven and Axelrod's Table 14 (p. 74), three other genera—*Hesperolinon* (Sharsmith, 1961), *Navarretia*, and *Streptanthus*—are richly endowed with narrow endemics. The general consensus is to consider these as *neo-endemics*: newly arisen "insular" species often demonstrating a pattern of adaptive radiation into specialized habitats. Although some woody genera (*Arctostaphylos* and *Ceanothus*) may conform to this model, it is the annual life-form that has epitomized the "insular," neo-endemic species model.

Raven and Axelrod provide a good discussion (pp. 73–83) of the possible modes of evolutionary diversification in these narrow endemics of late Tertiary and Quaternary origins. For some endemic annuals, the speciation-by-autogamy route may be postulated. The endemic populations are derived from nearby outbreeding congeners by the establishment on serpentine of preadapted autogamous genotypes as founders of a new line. They are usually not strikingly different from their outbreeding progenitors. Raven and Axelrod (p. 82) cite examples from *Clarkia, Camissonia, Mentzelia*, and others; these genera do contain confirmed or suspected serpentine endemics.

Harlan Lewis (1962) proposes another model for the origin of narrow endemics, with his compelling scheme of catastrophic selection resulting in saltational speciation; Raven (1964) then applies the Lewis model to the origin of narrow edaphic endemics. The scenario begins with a marginal population or one decimated by an unusually severe environmental stress, catastrophic to the bulk of the population. However, particular genotypes survive the catastrophe, bearing a new chromosomal rearrangement that fixes the preadapted genotype for the immediate future. Raven adds to this scenario the edaphic factor as the limiting environmental stress; serpentine would qualify as a prime milieu for stresses to act on plant populations.

So far, examples of saltational speciation are known only from *Clarkia*, and there is no certain case in this genus of a serpentine endemic with such an origin.[1] It remains to be seen whether the Lewis/Raven model can be fitted to the serpentine endemics in *Hesperolinon* and *Streptanthus*.

Speculation on the origin of narrowly restricted endemics need not be confined to the more unusual models discussed above. It is perfectly feasible in theory to plot origins via the allopatric speciation model of Mayr and others, or some variant of it (M.J.D. White, 1978). The allopatry model, coupled with any of several post-isolation events, could take the following form: In some period of Neogene time, serpentinites became available to colonization by preadapted elements of parent ancestral nonserpentine populations. Moderate spatial distance, coupled with extreme edaphic discontinuity, isolated populations; isolation permitted further divergence to produce distinct variants. Then Quaternary climatic changes wiped out the ancestral nonserpentine taxa.

A sympatric model can be hypothesized with equal facility. Spatial sympatry, coupled with extreme edaphic heterogeneity (serpentine outcrops in a "sea" of nonserpentine lithology), could promote divergence following establishment of preadapted genotypes on immediately adjacent serpentine. Depletion to extinction of nonserpentine biotypes would follow, to leave a serpentine population free to achieve species status.

The time factor is a critical ingredient in seeking explanations of evolutionary accommodation to serpentine. For California, Raven and Axelrod have made it clear that the serpentine of the Jurassic Franciscan formation has not been available for plants over anywhere near the time-span since their origin and emplacement. Colonization of serpentine has, in their view, only been possible during the Neogene (the late Pliocene to the Pleistocene). Wild and Bradshaw (1977) come to similar conclusions in an attempt to explain the accomodation of plants to serpentines of the Great Dyke in Rhodesia. The Great Dyke is Pre-Cambrian in age; were it continuously available throughout the intervening geological epochs, then surely it might have been a birthplace of early vascular plants, followed by later origins of primitive seed and flowering plants. But the Great Dyke is not a repository of a rich relictual flora; rather, its endemics appear to have been of recent origin. As for California, the late Tertiary–Pleistocene climatic changes must have prevented the persistence of earlier floras and fostered the evolution of new endemics.

STREPTANTHUS—A CASE HISTORY

The genus *Streptanthus* (Cruciferae) provides a provocative cast from which to create scenarios for testing various hypotheses on the origins of narrow endemism to serpentine.

1 The origin of *C. franciscana*, a serpentine endemic, once considered a type case for saltational speciation, has been reconsidered (Raven and Axelrod, 1978).

This New World genus consists of about 40 species, all confined to North America. About 30 species are restricted to the Californian Floristic Province, and 15 of them have affinities for serpentine. *Streptanthus* (including *Caulanthus* of Munz and Keck, 1959) can be divided into fairly distinct subgenera, some of which have clearly defined sections. No members of either subgenus *Eustreptanthus* (Arizona to Texas and Oklahoma) or *Caulanthus* (southern California and the intermountain states) show any affinity for serpentine, although some are edaphic specialists (limestone, gypsum, and alkali soils). This is to be expected, since there is no ultramafic rock in the areas of their occurrence. Subgenus *Paracaulanthus* (mostly southern and central California) has only one serpentine taxon, *S. amplexicaulis* Jeps. var. *barbarae* J.T. Howell. Subgenus Euclisia, with 16 species, is largely serpentinicolous, and 3 of its sections are exclusively on serpentine.

Streptanthus displays all degrees of restriction to serpentine, from edaphic (serpentine) races of otherwise nonserpentine taxa, through narrow endemics of no known affinity, to congeners in the genus. This graded display of serpentine addiction is discussed in more detail below.

A. Bodenvag taxa.

1. Taxa with both serpentine and nonserpentine races, but these races not warranting taxonomic recognition

Streptanthus glandulosus ssp. *glandulosus* and ssp. *secundus* (Greene) Kruckeberg, variants of the wide-ranging *S. glandulosus*, have been analyzed for their serpentine tolerance, crossability, and taxonomic status (Kruckeberg, 1951, 1954, 1957, 1958). This case seems to fit the ecotypic-differentiation model of edaphic tolerance. It either represents an early stage in adaptive radiation or may be an evolutionary accommodation to serpentine, now in equilibrium, of late Tertiary origin. Only with massive biotype depletion of nonserpentine races could it achieve the status of a serpentine endemic.

2. Species with named infraspecific taxa restricted to serpentine. Three species, each in different subgenera, have distinct varieties or subspecies endemic to serpentine.

Streptanthus amplexicaulis var. *barbarae* (subgenus Paracaulanthus) is on serpentine in the San Rafael Mountains of Santa Barbara County. Its nonserpentine var. *amplexicaulis* is concentrated in the San Gabriel and San Bernardino mountains of southern California; it is always on granitics there. Some intervening stations for var. *amplexicaulis*, as in the Lake Hughes area of northwestern Los Angeles County and on Frazier Mountain in eastern Ventura County, are also on granite. A singular site in the Lockwood Valley area (Thorne Meadow) is on a sterile talus of shale and conglomerate. These Ventura County plants are the closest to the var. *barbarae* of the Santa Barbara serpentines. Though different in substrate and in both vegetative and floral features, the geographic isolation of var. *barbarae* and var. *amplexicaulis* has not isolated the two genetically. They are fully interfertile in artificial crosses; F_1s have 95% good pollen and set viable seed (Kruckeberg, unpublished data).

Two variant taxa of the *S. glandulosus* complex (subgenus *Euclisia*, section *Euclisia*), *S. glandulosus* ssp. *pulchellus* and *S. albidus* Greene, are normally on serpentine; the Berkeley Hills localities of *S. albidus* are on both serpentine and nonserpentine rocks. Based on inter-crossing studies (Kruckeberg, 1957), I consider *pulchellus* a subspecies of *S. glandulosus*, while *S. albidus* is elevated to species rank (Kruckeberg, 1958). The *S. glandulosus* complex appears to be a system of populations in various stages of adaptive radiation, some achieving serpentine tolerance and genetic isolation.

The Sierran–north Coast Range species complex *S. tortuosus* (subgenus *Pleiocardia*, section *Tortuosi*) has one, or possibly two variants on serpentine. While the bulk of the Sierran *S. tortuosus* are annuals (var. *orbiculatus*) at mid to high elevations, usually on granitics, a very distinct biennial (or triennial, according to Jepson, 1936), *S. tortuosus* var. *optatus*, occurs on serpentine in the blue oak–digger pine belt. My unpublished studies show that the annual forms (var. *orbiculatus* (Greene) Hall) are intolerant to serpentine, while var. *optatus* is tolerant. The other variety, *oblongus* Jeps., is frequently found on serpentines of the Trinity River drainage in the north Coast Ranges. No interfertility tests with these distinctive variants have yet been conducted.

B. *Streptanthus* Species Endemic to Serpentine

Serpentine endemics, 11 of the 16 of subgenus *Euclisia*, are well set apart from other *Streptanthus* taxa. Three of them are very local in distribution. *S. niger* (section *Euclisia*) is confined to the tip of the Tiburon peninsula in Marin County; it stands as a morphological extreme manifestation of the *S. glandulosus* complex and is intersterile with its *glandulosus* congeners. *S. batrachopus*, restricted to one or two serpentine outcrops on Mt. Tamalpais, is also an extreme representative of its section, *Hesperides*. It is the smallest in flower and habit of any of its group, and a near obligate selfer. *S. insignis* ssp. *lyonii* Kruckeberg and Morrison is found on a very few local outcrops in the inner south Coast Ranges, in western Merced County. *S. insignis* ssp. *insignis* rather widespread on serpentine in the inner south Coast Ranges, is easily recognized by its distinctive purplish black "flag" (or "color spot") of terminal sterile flowers, and purplish fertile flowers. *S. insignis* var. *lyonii* has instead a pale yellow color spot, below which are yellow-green fertile flowers. Some western Merced County *lyonii* plants are bicolored, with yellow calyces and purplish petals. Plants of *lyonii* and *insignis* are artificially crossed with ease, and F_1 plants are highly fertile. This case seems to represent the fixation of a local, distinct biotype within the range of the parent species, with only a latent potential for speciation.

The other serpentine endemics in *Streptanthus* have more widespread distributions. Four of the 5 species in the section *Hesperides* are in the Coast Ranges. *S. breweri* occurs discontinuously from San Carlos Peak in San Benito County north to Lake and Colusa counties; its center of distribution is in Napa and Lake counties. *S. breweri* is replaced westward in western Lake, Sonoma and Mendocino counties by *S. barbiger*. And wholly within the range of both species is the distinctive *S. hesperidis*; it is often sympatric with *S. breweri*, though hybrids seem not to occur.

North of these 3 Hesperides species, a remarkable variant, *S. drepanoides* Kruckeberg and Morrison, assumes the annual *Streptanthus* niche on serpentine from western Colusa County north to Trinity County. All 5 *Hesperides* species—one a narrow endemic (*batrachopus*), and the others regional serpentine endemics—fit the model of Group III (Raven and Axelrod, 1978): narrow endemics of recent origin in rapidly diversifying genera. The question of how these "insular" species arise is probably unsolvable; their partially discontinuous distribution and patterns of replacement seem to suggest the possibility that one taxon gives rise to another in space and time. The other alternative, of a widespread ancestral taxon gestating all these derivative species, appears unlikely: there is no evidence for such a taxon.

Indeed, the *Hesperides* section is a distinct and coherent group with no close affinity to any other group in *Euclisia*—save one!

The other Euclisian section, *Biennes*, does have clear lines of affinity to section *Hesperides*. Its major distinctions are its biennial habit, growth form, and silique morphology; foliage and flowers are close matches to those of the *Hesperides* section. The two *Biennes* species, *S. brachiatus* and *S. morrisonii*, are obligate serpentine endemics in the Sonoma, Napa, and Lake county areas; they were discovered in the late 1940s by Freed Hoffman (1952). *S. brachiatus* is confined to the serpentine of the Big Geysers area in the Mayacamas range; *S. morrisonii* has two disjunct populations, one in northwestern Sonoma County (upper east Austin Creek) and one on the Napa–Lake counties line east of St. Helena Creek. Though there is little doubt about the close affinity between these *Biennes* species and those of the *Hesperides* group, reading the direction of the phylogeny is well nigh impossible. Both can coexist with annual members of the subgenus, so habitats are no more suited to the biennial than to the annual habit. Based on morphology alone, I can trace a possible sequence of species replacement: *S. breweri* gave rise to other annuals in *Hesperides*; one of these, *S. barbiger*, has foliar and floral attributes that make it a candidate intermediate between the annual and biennial groups.

The only Sierran endemic on serpentine, *S. polygaloides*, is also the most derivative, exceptional species in the entire genus. *S. polygaloides* occurs frequently but discontinuously along the narrow serpentine belt from Mariposa and Tuolumne counties to Butte County. As its species name suggests, its flowers look more like those of species in the genus *Polygala* than any other crucifer. The peculiar, zygomorphic flowers maintain their shape and size throughout its range, but vary appreciably in flower color—from yellow to deep rose. There is no trace of a living relative of this species, so completely unique is the plant. *S. polygaloides* Gray was considered by E.L. Greene (1904) as monotypic in the genus *Microsemia*; subsequent taxonomic reviews and floras have consistently placed the species in *Streptanthus*. Recently the genus *Microsemia* has been proposed for resurrection, based on geochemical studies. Reeves et al. (1981) have found that *S. polygaloides* is the only serpentine *Streptanthus* that can be classed as a "hyperaccumulator" of nickel. Other species tested take up only moderate amounts, but several populations of *S. polygaloides* all accumulated over 1000 ppm, some even up to 30,000 ppm. Reeves' group claims this to be the first known case of hyperaccumulator plants in North America. For this exceptional mineral uptake, in addition to its singular morphology, the authors suggest that the species be retained in *Microsemia*. The origin of this bizarre species is "an abominable mystery."

Two perennial species of *Streptanthus* (subgenus *Pleiocardia*, section *Cordati*) are regional serpentine endemics in the Klamath–Siskiyou Mountaina areas. Though their affinities to the widespread *S. cordatus* Nutt. seems substantial, both are unique entities morphologically. *S. cordatus* occurs throughout the arid West, mostly in desert mountain ranges. Two variants do occur on mafic to ultramafic rocks: *S. cordatus* var. *piutensis* J.T. Howell is on olivine gabbro in the Piute Mountains of Kern County, and a form recognized by E.L. Greene as *S. crassifolius* Greene is on serpentine in the Sierra in eastern Tehama County. But these are clearly variants of *S. cordatus*, whereas *S. barbatus* and *S. howellii* are very distinct and local taxa, exclusively on serpentine. Accommodation to serpentine has been complete for these two species. I envision them as having been recruited out of the gene pool of *S. cordatus* in the late Tertiary; the progenitor populations are now wholly disjunct (eastern California and southeast Oregon) from the serpentine endemics.

Particular subgenera of *Streptanthus*, then, can display a full range of steps toward absolute restriction to serpentine—from mere variants, within species not so confined, to highly singular taxa with little resemblance to living antecedents. Ecotypic differentiation, biotype depletion, allopatric (or parapatric) speciation, and perhaps even saltational speciation may be invoked to account for their origins. Studies in progress on phenetics and cladistics of the subgenus *Euclisia*, on chemotaxonomy of the entire genus (Rodman, Kruckeberg, and Al-Shebaz, 1981), and on interspecific crossing can lead to a more definitive interpretation of the origins of these fascinating serpentinophiles.

EXPLOITATION OF SERPENTINE AND OTHER ULTRAMAFICS AND EFFECTS ON PLANT LIFE

The natural diversity of our planet's biosphere is a natural resource, exploited or not. Since serpentines have a marked biotic effect in many parts of the world, these edaphic islands must be looked upon as significant segments of the worldwide fabric of diversity. Exploitation of serpentines as a resource for human use has occurred in a variety of ways, but can be readily categorized as either geophysical or biological. All such exploitation has had an effect on the biota. In California, both kinds of resource use have been under way ever since the arrival of western man. The effects on the vegetation and the flora have been substantial.

SERPENTINE AS A GEOPHYSICAL RESOURCE

Mining comes to mind easily, for ultramafics have had a long history of metal extraction. The first mines in California were probably quicksilver (mercury) mines, which are usually located on or adjacent to ultramafic deposits. The New Almaden Mine in Santa Clara County was in operation before the 1849 Gold Rush. Other heavy metals like chromium and nickel are extracted from outcrops of serpentinite or peridotite. One has only to look at maps of areas in California with ultramafic outcrops to discover ubiquitous mining activity. Names of mines, many now defunct, liberally dot the maps throughout the Jurassic Franciscan formation, where serpentines so commonly intrude.

Particularly rich mining regions in serpentine country include the north Coast Ranges (Mayacamas Mountains, the Klamath– Siskiyou region and nearly every county in between); the south Coast Ranges, especially San Luis Obispo, San Benito, and Santa Clara counties; and the Sierra Nevada foothill counties of Tuolumne and Mariposa.

Besides mercury and chromium, several other minerals have been extracted over the years from ultramafic rocks. Magnesite ($MgCO_3$) occurs in veins within serpentine and other ultramafics; between 1850 and 1965, upwards of $100,000,000 worth of this industrial mineral was extracted. Red Mountain in eastern Santa Clara County has been the site of the largest magnesite mine in California. New finds of asbestos in ultramafic rocks have made its extraction profitable in some parts of the state, notably western Fresno County. Talc, soapstone, and jadeite are also taken from ultramafics. It has not as yet proven profitable to extract nickel from ultramafics in California.

Vegetation or its constituent flora can often be used to locate and assess mineral concentrations of commercial value. The search for useful minerals with the aid of indicator plants or vegetation types has its own special niche in applied ecology; the term *geobotanical pros-pecting* is commonly applied to the practice, and chemical verification of mineral content in plant "guides" to ore deposits is called *biogeochemistry* (Cannon, 1960; Peterson, 1971; Brooks, 1972).

There has been little systematic use of serpentine floras and vegetation in California for the purpose of locating minerals. In a guide to the technique of geobotanical prospecting for California (Carlisle and Cleveland, 1958), the only example used was for molybdenum, found in other rock types. It would seem evident from what we now know about the unique occurrences of plants on most California serpentines that its flora could be a valuable guide to detecting mineralization. The recent report on the hyperaccumulation of nickel by *Strep-tanthus polygaloides*, cited earlier (Reeves et al., 1981), could trigger a flurry of geobotani-cal activity.

Mining can be construed as an abuse of the ultramafic outcrop, if one puts a premium on the unique biota of the serpentine. Loss of vegetation cover, extirpation of rare taxa, exploi-tation of sparse forest trees for mine timbers, etc., have materially altered the plant life of serpentines over the more than 100 years of mineral extraction. I want to come back to the question of conservation of serpentine ecosystems after examining other "abuses," where exploitation has been put ahead of preservation.

The most massive alteration in store for some larger California serpentine exposures can result from geothermal power development. All the present [1984] geothermal power tech-nology in California is centered around the Geysers area of Sonoma and Lake counties, in the Mayacamas Mountains. Serpentine outcrops, often many hectares in extent, are intimately associated with the exploitation and construction efforts associated with tapping steam power.

Geothermal power development requires roads for access by heavy equipment, installa-tion of devices for tapping and conducting the steam, etc. All such activities have a direct impact on serpentine vegetation. The removal of woody plant material affects the serpentine chaparral and woodland by destroying samples of its unique woody species (*Cupressus sar-gentii, Quercus durata, Ceanothus jepsonii*, etc.), as well as other elements of the serpentine flora. The impact on populations of the rare local *Streptanthus brachiatus* has been of special concern to botanists and, perforce, to the developers of the energy projects.

Major serpentine areas in the state can be looked at as irreplaceable watershed systems. The highly fractured and faulted nature of ultramafic outcrops fosters the occurrence of springs, seeps, and other sources of continuous water flow. Where creeks in nearby nonser-pentine areas are dry in the summer, they tend to flow all year round on ultramafics. Heavy rains can cause some sheet erosion and mass wasting on even undisturbed serpentine barrens; but disturbance, especially excavation for roads and other installations, severely enhances the erosion potential on serpentines.

BIOLOGICAL EXPLOITATION OF SERPENTINE HABITATS

Agriculture

Serpentine soils are deceptive; in alluvial areas they can be deep and heavy-textured and appear as though they would be fertile. Yet most attempts to grow crops (barley, alfalfa, etc., for hay) have been unsuccessful. Where the alluvial soils are predominantly of serpentine origin, the unfavorable base status (high magnesium and low calcium) works against successful cropping. Heavy applications of gypsum ($CaSO_4 \cdot H_2O$)to increase the calcium status must be repeated. Even at 8 tons per acre of gypsum, only the top 6 inches of a Venado or Maxwell series soil had sufficiently improved calcium status to promote satisfactory growth of hay forage plants (Martin et al., 1953).

In a recent study (Jones et al., 1977), 23 residual serpentine soils of the Henneke series responded well to heavy applications of phosphorus and sulfur. Molybdenum deficiency (Walker, 1948) and possible heavy metal toxicities (nickel and chromium) also can work against growing crop plants on alluvial soils from nearby upland serpentine sites.

Livestock have been on the serpentine trail for many decades. Upland as well as alluvial flats have served as marginal pasturage, probably ever since Spanish land-grant days. Low productivity of serpentine vegetation, the paucity of palatable species, and the often steep, barren and unstable terrain afford only a marginally low carrying capacity. And of course cattle can "crop" narrow endemics just as easily as they can graze on commoner species; substantial cattle impact can precipitate erosion problems, too.

Forestry

Foresters have generally considered stands of forest species on ultramafics as occurring on sites too low for sustained yield management. "Serpentine areas within forest zones are classed as noncommercial and are not included in the timber base of the local management units" (Rai et al., 1970). This has not precluded the harvesting of trees on many serpentine areas in California. Although old-growth stands are often thinly stocked, size classes with older trees often include specimens of great girth and biomass. Old-growth trees on serpentine, however, have attained their maturity at much slower growth rates; annual increments are low. Once cut, the old, "over-mature" trees may not be replaced by old second-growth, merchantable trees in less than 150 years. Foresters recognize the slow growth on serpentine: "Incense cedar is also found in isolated spots in the Coast Ranges growing on soils developed from serpentine. Though it grows better than other conifers on these soils, its growth is considerably less than on the other soil types" (Fowells, 1965, p. 243).

Serpentine areas with tree species of commercial value are mainly in northwestern California (Mendocino County north to Del Norte and Siskiyou counties). Here ultramafics support Jeffrey pine, incense cedar, and Douglas fir. I know of no serpentine areas with coast redwood, sugar pine, yellow pine, or true firs in commercially usable stands.

No doubt any available tree species on serpentine have been cut down for some local need; even digger pine or Sargent cypress are harvested for fenceposts or firewood. The early days of quicksilver extraction saw many nearby areas nearly denuded of trees to satisfy the insatiable need for furnace wood: "Quicksilver furnaces are great consumers of wood, and even those mines which are located in well-timbered regions find the cost of their fuel steadily increasing" (Bradley, 1918, p. 15). Further, mine timbers were often secured from nearby stands. For instance, we read in Bradley's account of the famous New Idria mine (San Benito County) that "at the portal of #2 level there was formerly a steam-driven sawmill of 2000 b.m. ft. daily capacity. Round timbers were used—both pine and—cedar—some being cut on the land owned by the company . . . " (p. 114). Very likely these timbers were taken from Jeffrey pine and incense cedar, both local indicators of serpentine in the New Idria area.

Timbered serpentine habitats are mainly in the higher rainfall areas of the state—the north Coast Ranges and the Klamath–Siskiyou region. Timber harvest, like mining or geothermal operations, creates serious erosion hazards here. Serpentine soils are so very easily prone to slump and then come downslope in massive sheets of mud and rock; timber harvest merely exacerbates this natural tendency on steep serpentine slopes.

Wildlife and Watershed Resources on Serpentines

Some recreational use of serpentine areas seems inevitable. The open terrain, often devoid of obstructing vegetation, has been inviting to pleasure-seekers on all-terrain vehicles and motorbikes. The New Idria serpentines have received the brunt of such assaults—those wide expanses of serpentine "desert" are the nearest topographic matches for dunes. Unfortunately, rare plants do occur on the serpentine barrens: *Layia discoidea*, one of the rarest plants in California, gets assaulted annually. The Bureau of Land Management has been persuaded to post signs designating the surroundings as a Natural Area. There is also a health hazard associated with ATV-riding on ultramafics: the presence of asbestos fibers in the dust presents a danger akin to tobacco smoking—"Inhaling serpentine dust may be hazardous to your health" might be an appropriate sign at New Idria.

The serpentine areas in the Sierra foothills, especially near Chinese Camp, have also created a confrontation scenario between ATV users and conservationists.

In central and northern California, hunting is a time-honored use of serpentine habitats. And this brings me to the most benign use of serpentine areas: "managed" as wildlife habitats. Deer, rodents, birds, and other wildlife utilize the vegetation of serpentine for forage, cover, breeding, etc. Though there are no known serpentine races of vertebrates, many species of the California vertebrate diversity can be found on serpentines.

The other benign use of serpentine habitat is of course as watershed. Not only is surface runoff into natural and manmade catchments substantial on California serpentines; the very nature of the parent material—faulted and fractured metamorphic and igneous ultramafic rock—serves to store meteoric water in the water table. Year-round live streams and many springs are ample evidence of the serpentine terrain's contribution to the water resources of the state. Indeed, all of the Soil Conservation Service's soil surveys that include serpentine soils of upland locations are unanimous in their recommendation that the highest and best use of these habitats is for the wildlife and water resources.

From the above, it is apparent that human activity on ultramafics has been substantial over the years, though hardly on the scale witnessed by the massive modifications of valley alluvial soils in California. All over the state, efforts are under way to preserve samples of natural habitats. With the unique flora of serpentine soils, the question of preservation clearly emerges. In the next section I explore the problems encountered by custodians of serpentine landscapes, who must balance human uses with preservation of habitats.

LAND MANAGEMENT AND CONSERVATION ON ULTRAMAFICS

The hand of man has writ large on the many and varied landscapes of California. Regions with ultramafic rocks have witnessed their share of human intrusion. Resource extraction—mining, geothermal power development, and timber—has made the major inroads on the geophysical and biological substance of serpentine country. I have recited above, examples of the variety of human exploitation and their particular consequences. Present and future uses necessitate guidelines for management of serpentine areas, both for their resource potential and for their intrinsic worth as unique places in the natural world.

Mining, including geothermal power extraction, will continue. Heavy metals (nickel, chromium, mercury, etc.), magnesite, and asbestos are still to be had, though cost-benefit considerations may delay their exploitation until shortages—or wartime restriction of other sources—urge their recovery. Public and private land managers of ultramafic deposits are now forced to consider the environmental impact of development activities. The case of serpentine areas is particularly instructive now, because of state and federal insistence on recognition of rare or endangered species.[1] Since narrow endemics are frequently encountered on serpentines, environmental-impact studies have tended to focus on ways of mitigating such impacts. But other consequences of development can be serious: damage to watershed (excessive runoff, erosion, and flooding) and to wildlife habitat can also be substantial impacts. Exploitation for geophysical resources modifies whole plant communities on serpentine; though these may not contain rare flora, the plant associations (serpentine chaparral, woodland, or grassland) are unique assemblages of plants.

As the more productive agricultural and forest lands are increasingly exploited, there will be a search for ways to make marginal lands more productive. Upland as well as basin (alluvial) serpentine areas will be targeted for improvement. Some efforts are already pointed in these directions. Rangeland improvement on chaparral sites in California aims at replacing low-palatability species (mostly woody chaparral plants) with forage grasses. Serpentine chaparral and serpentine woodland have been included among those ecosystems that can be manipulated for improved forage yields. Unpublished results from such efforts by the Hopland Field Station of the University of California at Davis have not been promising. Some clovers and grasses have been established on serpentines following nutrient supplements (M.B. Jones, personal communication). Yet deficiencies in phosphorus and nitrogen were

---

1 A test of California's rare-plant laws may be imminent at this writing. Planning is under way for a massive nickel-extraction project—on ultramafics, of course—in the Gasquet area of Del Norte County. Some of the rarest of California serpentine endemics, as well as outstanding examples of mesic serpentine floras, occur here. The California Native Plant Society and others, who hold that plants have rights too, will have a battle on their hands.

evident, although the clovers did improve the nitrogen status. The major problem with such palatable species has been predation by wildlife; unprotected plots were grazed bare.

Two studies by Koenigs et al. (1982) were specifically undertaken to provide vegetation and chemical data so that land managers could better assess the success of replacing serpentine vegetation with forage grasses. They suggest that annual grasses have a better chance for success in "non-cypress" serpentine vegetation and within the "non-cypress" stands, the chamise (*Adenostoma fasciculatum*) and silk-tassel bush (*Garrya congdonii*) sites would better support grasses because of the more favorable Ca status.

The continued attrition of undisturbed (or relatively pristine) serpentine areas in California inevitably leads to problems of habitat conservation. It can be taken as a "given" that by preservation of relatively undisturbed serpentine areas, both the plant communities and their rare biota can be retained as significant elements in California's heritage of natural diversity. Preservation is a complex objective, with political, economic, and ecological intricacies. From the ecological standpoint, can serpentine areas be preserved by simply leaving them alone? An answer to this question depends on whether serpentine vegetation is a relatively steady-state ecosystem with stabilized diversity and biomass. Proctor and Woodell (1975) take up this question in the sense of the Clementsian climax concept. They side with Whittaker (1960), who found no evidence for seral status of serpentine vegetation in the Siskiyou region. Proctor and Woodell (p. 269) further assert that this view is "true for many other [serpentine] areas also." I concur, and thus would predict that custodianship of samples of serpentine vegetation might need only basic protection against wildfire and excessive human impact. For short- and middle-term conservation of ecosystems, methodologies to maintain seral status will not be necessary. Only if climatic or biotic influences perturb the steady-state nature of serpentine habitats will land managers need to devise measures to preserve the desired (− status quo) plant life.

Preservation of samples of serpentine habitat as natural areas has been sporadic. Most serpentine vegetation in remote areas begets de facto preservation through neglect. Areas more accessible may or may not have become grossly altered through disturbance. Often serpentine habitats manage to escape most human activity due to their low productivity, barren appearance, or physical instability.

The widespread and discontinuous distribution of serpentinites, outcropping from southern California to the Oregon border, from sea level to upper montane altitudes, from the coast to the Sierra means that ownership is highly variable. Private ownership may be residential, agricultural, industrial, or nondeveloped land. Serpentines in the public domain may be under local, county, state, or federal jurisdiction. Preservation of serpentine habitats thus takes on a wide array of manifestations, from active conservation to incidental custodianship or even benign neglect. Table 6 summarizes some of the kinds of preservation in the state, as well as particular examples of preserves.

The Rare and Endangered Species Act of 1973 finally embraced plants a few years later. Progressive refinement of the original Smithsonian list of endangered, threatened, and extinct plants (Ayensu and DeFilipps, 1978) is still under way. A number of serpentine endemics have appeared on various lists for the California flora. The most comprehensive list of rare and endangered plants for California, prepared by the Rare Plant Project Group of the California Native Plant Society (Smith et al., 1980), is now the standard reference list for the state. Both state and federal lists (see Table 7) will be monitored and brought up-to-date in

Table 6. Natural areas on ultramafic substrates

Locality	Jurisdiction/Management	Habitat features	Threats/conflicts
Private land			
1. Howell Wildflower Garden, Tiburon Peninsula, Marin County	Nature Conservancy and local citizens	Serpentine grassland; rare taxa: *Streptanthus niger*, *Calochortus tiburonensis*, *Castilleja neglecta*	Land development
2. Skyline Blvd., San Mateo County	Homeowners; real estate developers	Serpentine grassland and chaparral; rare *Fritillaria*	Land development
3. Metcalf Canyon, Santa Clara County	Farmland	Serpentine grassland; *Streptanthus albidus*	Agricultural and land development
4. The Cedars-Big Slide area, Big Austin Creek, Sonoma County	Santa Rosa Girl Scout Council (?)	*Cupressus sargentii* pygmy forest; *Streptanthus morrisonii*	Mining?
5. Geysers area, Mayacamas Mts., Sonoma and Lake counties	Mining claims; geothermal claims	Outstanding serpentine chaparral and woodland; *Streptanthus brachiatus*	Mining, geothermal development, and roads
Local and regional government			
1. Crystal Springs Reservoir, San Mateo County	Regional water supply	Serpentine grassland and chaparral; rare taxa: *Acanthomintha obovata* spp. *duttonii*, *Cirsium fontinale*	None, if *status quo* persists
2. Carson Ridge area of Mt. Tamalpais watershed, Marin County	Regional water supply	Serpentine chaparral and woodland; *Cupressus sargentii*, *Streptanthus batrachopus*	None, if *status quo* persists

70

Location	Ownership/Status	Vegetation	Use
3. Sunol County Park, Alameda County	Alameda County	Serpentine vegetation	Recreational
4. Lake Hennessey Recreation Area, Napa County	City of Napa	Serpentine chaparral at Conn Dam	Recreational
5. Milliken Recreation Area, Napa County	City of Napa	Serpentine chaparral?	Recreational
6. City of San Francisco	City Parks	Remnants of serpentine vegetation in parks/open space	

State parks

Location	Ownership/Status	Vegetation	Use
1. Mt. Tamalpais, Marin County: Bootjack Camp; summit area; Rifle Camp	Recreation, wildlife, flora	Serpentine chaparral and woodland; *Streptanthus glandulosus pulchellus*, *S. batrachopus*	Recreation, development
2. Mt. Diablo, Contra Costa County	Recreation, wildlife, flora	Serpentine grassland: *Streptanthus albidus*; serpentine chaparral	Recreation, development
3. Robert Louis Stevenson (undeveloped), Mt. St. Helena area, Napa-Lake-Sonoma counties	Undeveloped?	Serpentine chaparral and woodland; *Erythronium helenae*	Unknown

Other state lands

Location	Ownership/Status	Vegetation	Use
1. Boggs Mt. State Forest, Lake County	Undeveloped?		
2. Austin Creek State Recreation Area, Sonoma County	Undeveloped?		

Table 6. (cont.)

Federal lands—Coast Ranges

Forest Service

Los Padres National Forest

1. Cachuma Saddle, San Rafael Range, Santa Barbara County	Recreation, watershed	Chaparral; *Streptanthus amplexicaulis barbarae*	Minimal
2. Figueroa Mtn., San Rafael Range, Santa Barbara County	Recreation, watershed	Chaparral, serpentine woodland; *Cupressus sargentii*, *Streptanthus amplexicaulis barbarae*	Minimal
3. Cuesta Pass area, Santa Lucia Range, San Luis Obispo County	Recreation, watershed	Chaparral and woodland; *Cupressus sargentii*, *Arctostaphylos obispoensis*	
4. Chew's Ridge area, Santa Lucia Range, Monterey County	Recreation, watershed	Grassland and woodland; *Calochortus invenustus*, *Aspidotis densa?*	
5. Pine Ridge area, Santa Lucia Range, Monterey County	Recreation, watershed	*Aspidotis densa* in serpentine grassland	

Mendocino National Forest

Many serpentine localities from Lake County to Yolla Bollys	Mostly multiple use		
6. Snow Mtn., north Lake County	Watershed, timber	Serpentine woodland and chaparral	Other resources
7. Frenzel Creek, Colusa County	Research natural area (federal system)	Serpentine woodland and chaparral; *Streptanthus* (undescr. sp.)	None

72

Trinity National Forest

Many serpentine localities from Yolla Bollys to Trinity Alps, mostly multiple use — Mostly multiple use

Locality	Use	Vegetation	Threats
8. Dubakella Mtn., Trinity County	Multiple use	Montane serpentine woodland: Jeffrey pine-incense cedar; serpentine endemic species	Other uses
9. Middle Eel-Yolla Bolly Wilderness, Trinity and Tehama counties	Wilderness	Same as No. 8.	Probably none

Six Rivers National Forest

Many serpentine localities from southwest Trinity County to Del Norte County

Locality	Use	Vegetation	Threats
10. Horse Mtn., Humboldt County	Recreation	Jeffrey pine-*Quercus vaccinifolia* with serpentine endemics: *Antennaria suffrutescens*, etc.	Other uses
11. Red Mtn. (may not be in NF), Mendocino County	Multiple use	Serpentine woodland; rare taxa	Mining, logging
12. Gasquet area in Smith River Canyon, Del Norte County	Multiple use	Serpentine woodland; rare taxa: *Arenaria howellii*, *Streptanthus howellii*	Mining, logging, development
13. Low Divide area, Del Norte County	Multiple use	Serpentine woodland; many rare taxa	Other uses

Table 6. (cont.)

Klamath (Siskiyou?)
National Forest

Many serpentine localities
from Trinity Alps to Oregon
border

Location	Use	Vegetation	Status
14. Youngs Valley-Preston Peak-Sanger Peak area, Siskiyou County	Multiple use	Serpentine woodlands and barrens; rare taxa	Other uses
15. Happy Camp-Klamath River area, Siskiyou County	Multiple use	Serpentine woodland, chaparral and barrens; rare taxa	Other uses
16. Cook and Green Pass area, Siskiyou County	Multiple use	Woodland and barrens; rare taxa	Other uses
17. Marble Mt. Wilderness, Siskiyou County	Wilderness; recreation	Montane serpentine vegetation likely	Low impact
18. Salmon-Trinity Alps Primitive Area, Trinity County	Recreation	Montane serpentine vegetation	Low impact
Shasta National Forest			
19. Scott Mts., Trinity County	Multiple use	Jeffrey pine serpentines; rare taxa	Other uses
20. Trinity River canyon, Trinity County	Multiple use	Serpentine chaparral and woodland	Other uses
21. Mt. Eddy, Shasta County	Multiple use; mostly recreation	Montane forest and subalpine on peridotite; rare taxa	Other uses

74

Bureau of Land Management, Bureau of Reclamation

1. San Benito Mtn., San Carlos Peak area, San Benito County	BLM, recreation	Isolated stands of Jeffrey pine-incense cedar; rare taxa	ATV activity
2. Lake Berryessa, Napa County	Bur. Rec.; recreation	Serpentine chaparral with endemics and indicators	Boating, camping, ATV

U.S. Army

1. Presidio, San Francisco	Recreation; developed areas	Endangered serpentine flora: *Clarkia franciscana*; *Arctostaphylos hookeri franciscana*	Protected?

Federal lands—Sierra Nevada

U.S. Forest Service
No special protection in any of National Forests: Plumas, El Dorado, etc.

1. Feather River Canyon, Plumas National Forest, Butte and Plumas counties	Multiple use	Mostly forested serpentines; rare taxa: *Streptanthus polygaloides*, *Cupressus macnabiana*	Other uses

U.S. Park Service
No major serpentine areas in this jurisdiction (most serpentine areas in the Sierra Nevada are in private ownership)

University preserve

1. Jasper Ridge, San Mateo County Stanford University		serpentine grassland	Protected

future years.[2] Development projects must comply with the California Environmental Protection Act by determining if listed rare plants will be disturbed by any significant alteration of the habitat. The most notable case where a rare serpentine plant has demanded recognition, study, and protection during development is *Streptanthus brachiatus*, endemic to serpentine in the Geysers geothermal belt (Lake and Sonoma counties). In the course of environmental assessment, ecologists working on the project have thoroughly mapped the distribution of the taxon (Neilson, 1977), and in addition have discovered significant variants of this remarkable serpentine endemic.

2   This activity is managed in California by the California Natural Diversity Data Base, Sacramento 95814.

Table 7. Serpentine plants designated as rare and endangered by state and federal agencies[1,2]

Taxon	Locality	Federal listed	List under review[1]	CNPS list, 1974[2]	Removed from 1980 CNPS list
Acanthomintha obovata ssp. duttonii	San Mateo Co.		x	x	
Allium hoffmanii	North Coast Ranges		x	x	
Antennaria suffrutescens	North Coast Ranges			x	x
Arabis constancei	Plumas County, Sierra Nevada		x	x	
A. macdonaldiana	Mendocino Co.	x	x	x	
A. serpentinicola	Siskiyou Co., Klamath Ranges		x	x	
Arctostaphylos bakeri	Sonoma Co.		x	x	
A. hookeri ssp. franciscana	Presidio, San Francisco		x	x	
A. hookeri ssp. ravenii	Presidio, San Francisco	x	x	x	
A. montana	Mt. Tamalpais, Marin Co.		x	x	
A. stanfordiana ssp. hispidula	Del Norte Co.		x	x	x
Arenaria howellii	Del Norte Co.		x	x	x
A. rosei	Trinity Co.		x	x	
Benitoa occidentalis	Inner south Coast Ranges		x	x	
Brodiaea pallida	Chinese Camp, Tuolumne Co.		x	x	
Calochortus greenei	Siskiyou Co.		x	x	
C. obispoensis	San Luis Obispo Co.		x	x	
C. tiburonensis	Tiburon, Marin Co.		x	x	
Camissonia benitensis	San Benito Co.		x	x	
Carex obispoensis	San Luis Obispo Co.		x	x	

Table 7. (cont.)

Species	Locality			
Castilleja brevilobata	Del Norte Co.	x		
C. neglecta	Tiburon, Marin Co.		x	x
Caulanthus amplexicaulis ssp. *barbarae*	Santa Barbara Co.		x	x
Ceanothus ferrisae	Santa Clara and Santa Cruz Cos.		x	x
Chlorogalum grandiflorum	Tuolumne Co.		x	x
Chorizanthe breweri	San Luis Obispo Co.		x	x
Cirsium fontinale var. *fontinale*	San Mateo Co.		x	x
C. fontinale var. *obispoense*	San Luis Obispo Co.		x	x
C. campylon	Santa Clara Co.		x	x
C. hydrophyllum var. *vaseyi*	Marin Co.		x	x
Clarkia franciscana	Presidio, San Francisco		x	x
Cordylanthus nidularis	Contra Costa Co.		x	x
C. tenuis ssp. *capillaris*	Sonoma Co.			x
Coreopsis hamiltonii	Inner south Coast Ranges (serp?)		x	x
Cupressus nevadensis	Kern Co. (mafic rock)	x		x
Darlingtonia californica	North Coast Ranges; Sierra Nevada	x		x
Dicentra formosa ssp. *oregona*	Del Norte Co.		x	x
Dudleya bettinae	San Luis Obispo Co.		x	x
Epilobium nivium	Lake and Mendocino Cos.		x	x
Eriogonum alpinum	Trinity Co.		x	x
E. kelloggii	Mendocino Co.		x	x
E. pendulum	Del Norte Co. (serp?)		x	x
E. siskiyouense	Siskiyou Co.	x	x	x

Species	Distribution				
E. vestitum	San Benito Co.			×	×
Eriophyllum latilobum	San Mateo Co.		×	×	×
Erysimum franciscanum	San Mateo to Marin Cos.		×	×	×
Erythronium tuolumnense	Tuolumne Co. (serp?)		×	×	
Fremontodendron decumbens	El Dorado Co. (olivine gabbro)		×	×	
Fritillaria falcata	Inner south Coast Ranges		×	×	
F. viridea	San Benito and Monterey Cos.		×	×	
Galium hardhamiae	Monterey and San Luis Obispo Cos.		×	×	
Haplopappus ophiditis	North Coast Ranges		×	×	
Helianthus bolanderi exilis	Napa and Lake Cos.		×	×	
Hemizonia halliana	San Benito, Monterey, and San Luis Obispo Cos.	×	×	×	
Hesperolinon adenophyllum	North Coast Ranges	×	×	×	
H. bicarpellatum	Napa and Lake Cos.		×	×	
H. breweri	North Coast Ranges: Contra Costa to Yolo Cos.		×	×	
H. congestum	Bay Area		×	×	
H. didymocarpum	Lake Co.		×	×	
H. drymarioides	Inner north Coast Ranges		×	×	
Laya discoidea	San Benito Co.		×	×	
Lilium vollmeri	North Coast Ranges (serp?)		×	×	
L. wigginsii	North Coast Ranges (serp?)		×	×	
Lomatium congdoni	Tuolumne and Mariposa Cos.		×	×	
L. howellii	De Norte Co.	×	×	×	
L. peckianum	Siskiyou Co. (serp?)		×	×	

Table 7. (cont.)

Species	Distribution		
Lupinus spectabilis	Mariposa and Tuolumne Cos.		x
Monardella benitensis	Monterey, San Benito, and San Luis Obispo Cos.	x	x
Perideridia leptocarpa	Siskiyou Co.	x	x
Phacelia dalesiana	Trinity Co.		x
P. greenei	Siskiyou Co.		x
P. phacelioides	Contra Costa to Santa Clara Cos.		x
Raillardella pringlei	Siskiyou Co.		x
Sanicula peckiana	Del Norte Co.	x	x
S. tracyi	Trinity and Humboldt Cos.		x
Sedum albomarginatum	Butte and Plumas Cos.		x
S. laxum ssp. *eastwoodiae*	Glenn, Lake, and Mendocino Cos.		x
Senecio clevelandii var. *heterophyllus*	Tuolumne Co.	x	x
Sidalcea hickmanii ssp. *anomala*	San Luis Obispo Co.		x
S. hickmanii ssp. *viridis*	Marin Co.		x
Streptanthus albidus ssp. *albidus*	Santa Clara Co.		x
S. batrachopus	Marin Co.		x
S. brachiatus	Sonoma and Lake Cos.		x
S. howellii	Del Norte Co.??		x
S. morrisonii ssp. *elatus*	Lake and Napa Cos.		x
S. niger	Marin Co.		x
Tauschia howellii	Del Norte and Siskiyou Cos.		x
Thlaspi montanum var. *siskiyouense*	Del Norte and Siskiyou Cos.?	x	x

80

Veronica copelandii   Trinity and Siskiyou Cos.

Wyethia reticulata   El Dorado Co. (serp?)

[1] U.S. Federal Register, Dec. 15, 1980, pt. LV.

[2] Powell, 1974.

[3] Smith et al., 1980.

SUMMARY

1. The great diversity of California's natural environments includes over 1100 sqare miles of ultramafic (serpentine and other ferromagnesian rock) outcrops. Species composition and vegetation cover are profoundly affected by the serpentine rock and their derived soils.

2. Most early California botanists failed to record their impressions of the remarkable association of unusual plant life with serpentine soils. Apart from William Brewer (1860s), none of the luminaries in California botany (Eastwood, Greene, Heller, Jepson, Jones, etc.) made mention of the serpentine substrate and its unique flora. Herbert Mason in the 1940s was the first to give serpentine plant life proper recognition. The state floras by Abrams and by Munz and Keck often record the preferences of certain plants for serpentine; Jepson's two major works on the flora do not refer to the serpentine habitat.

3. Soil scientists and physiologists (Gordon, Lipman, Robinson, etc.) attempted to relate the chemical nature of serpentine soils to plant growth in the 1920–1940 period.

4. California geologists recorded the presence of outcrops beginning early in the 19th century (HMS Blossom expedition, Whitney's geological survey of the state). Economic geologists of the early 20th century recognized the indicator value of serpentine outcrops and their sparse or barren look in searching for quicksilver and other minerals.

5. Serpentine outcrops are largely distributed in cismontane California: in the foothills of the Sierra Nevada and in the Coast Ranges from San Luis Obispo County north to the Oregon border. Ultramafics are frequent constituent rocks in the Jurassic Franciscan formation of California.

6. Ultramafic rocks outcrop in California as either igneous (peridotite, dunite, or harzburgite, in the north) or metamorphics (serpentinite rock). Serpentinite is the more common rock type in both the Sierra Nevada and the Coast Ranges.

7. The ultramafic rocks consist of iron–magnesium silicates; olivine, chromite, and pyroxene are the major mineral forms. Serpentinite consists mainly of the minerals chrysotile (asbestos), lizardite, and antigorite; iron and magnesium are also major elements.

8. Ultramafics may have been available (exposed) for plant colonization at least through parts of the Tertiary, though most outcrops may have become exposed only in the late Tertiary (Miocene to Pleistocene).

9. The major occurrences of the ultramafics are in the south Coast Ranges, the San Francisco Bay Region, and the north Coast Ranges west of the Central Valley; in the Sierra Nevada, outcrops range from Tulare to Plumas counties.

10. Serpentine and other ultramafics weather to soils rich in magnesium, iron, and silicates. Montmorillonite is the typical clay mineral of serpentine soils.

11. Several soil series derived from serpentinite have been recognized for the state; the Henneke, Dubakella, and Montara series are the most common residual (primary or upland)

types. Four basin (alluvial) soils, derived from adjacent upland serpentine soils, have been named. The upland soils are usually shallow and stony, with little profile development; they are usually identified as skeletal or azonal soils.

12.   Unique features of the cation status and mineral nutrition of serpentine soils include high values for exchangeable magnesium and exceptionally low values for exchangeable calcium. The soils are also typically deficient in nitrogen and phosphorus; molybdenum deficiency is also known for California serpentine soils. The heavy metals—iron, chromium, and nickel—can attain high values in both soil and plant tissues. One instance of hyperaccumulation (more than 1000 ppm in tissue) of nickel by a native serpentine endemic, *Streptanthus polygaloides*, has been recently reported.

13.   The causes of the "serpentine barren" syndrome have been traced to magnesium toxicity, imbalance of calcium and magnesium, heavy metal toxicity, and low nitrogen levels. The single-factor explanation is abandoned here in favor of a multiple-factor one: the interaction of stresses in chemical composition of the soils, the physical nature of the habitat, and the biological consequences of inadequate recycling of nutrients may afford a more complete, holocoenotic explanation.

14.   Physiological responses to serpentine soils must be characterized on a species-to-species basis. Different physiological "strategies" are employed by different species to tolerate the serpentine conditions.

15.   Vegetation on California serpentines is usually markedly different in physiognomy and species composition from that on adjacent nonserpentine vegetation. In California, the change in vegetation type (and life-form) across a serpentine contact usually involves a shift from oak savannah (e.g., blue oak and digger pine) to serpentine "hard" chaparral, from Douglas fir and hardwood to pine and cypress, or from chaparral to sparse grassland.

16.   The most common vegetation type on serpentine is a kind of hard chaparral characterized by both serpentine-endemic sclerophylls (*Quercus durata, Ceanothus jepsonii,* and *Garrya congdonii*) and nonendemic species. In extreme northwestern California (the Klamath–Siskiyou area), Jeffrey pine–incense cedar open woodland typically occurs on ultramafics, replacing the nonserpentine forests of yellow pine, Douglas fir, redwood, or mixed conifers and hardwoods.

17.   The challenge of serpentine has produced four major categories of taxonomic and evolutionary responses in the California flora: 1. Taxa endemic to serpentine, either local or widespread; 2. taxa, called here local or regional indicators, that are largely confined to serpentine in only parts of their geographic ranges; 3. taxa, variously called bodenvag, ubiquist, or indifferent, that freely range on and off serpentine (though some may be ecotypically differentiated); 4. taxa that "avoid" or are excluded from serpentine.

18.   Serpentine endemics include 215 known species, subspecies, and varieties of the native flora that are restricted wholly or in large part to serpentine soils. These include all major life-forms of flowering plants (monocots and dicots), as well as some ferns. Some examples are: trees: *Cupressus sargentii* and *C. macnabiana*; shrubs: *Quercus durata, Ceanothus jepsonii,* and *Arctostaphylos hookeri* var. *franciscana*; herbaceous perennials: *Calochortus tiburonensis,* and *Galium hardhamiae*; many annuals: *Streptanthus* (13 species of subgenus Euclisia)—e.g., *S. breweri, S. brachiatus, S. insignis*; 8 species of *Hesperolinon*—e.g., *H. adenophyllum* and *H. disjunctum*; and a variety of annual composites—e.g., *Layia discoidea, Benitoa occidentalis,* and *Lessingia ramulosa*.

19.   There are two intergradient categories of serpentine indicators; all include taxa that

locally or regionally characterize serpentine habitats, but are common on other substrates: 1. taxa that are faithful to serpentine in one segment of their total range, geographic and altitudinal—e.g., *Pinus jeffreyi*, *Calocedrus decurrens*, and *Darlingtonia californica*; 2. taxa that occur on serpentine throughout most of their total range, as recruits from local xeric nonserpentine habitats—e.g., *Pinus sabiniana*, *P. attenuata*, *Ceanothus cuneatus*, and *Adenostoma fasciculatum*.

20. Bodenvag or indifferent taxa are elements of a local flora that are found on both serpentine and adjacent nonserpentine substrates; some may be ecotypically differentiated into serpentine-tolerant and -intolerant races. Some examples are: *Achillea lanulosa*, *Gilia capitata*, *Salvia columbariae*, *Pteridium aquilinum*, *Adiantum pedatum*, several introduced annuals (e.g., *Bromus mollis*, *Hypochaeris radicata*), and many other native taxa.

21. Many plants on nearby normal (zonal), nonserpentine soils do not appear on serpentine. Some examples are: trees: *Pinus ponderosa*, *Abies concolor*, and *Quercus agrifolia*; shrubs: *Cercocarpus betuloides*, several species of *Ribes*, *Rubus*, *Ceanothus*, and *Arctostaphylos*; herbs: *Amsinckia*, *Draba*, *Erigeron*, *Heuchera*, *Lotus*, *Penstemon*, *Potentilla*, *Ranunculus*, *Saxifraga*, *Solanum*, and *Trifolium*. Lack of genetic preadaptation is hypothesized as the cause of this intolerance to serpentine.

22. Manifestations of the serpentine response vary geographically, latitudinally, and altitudinally. Xeric taxa predominate in the south and central Coast Ranges. The Klamath–Siskiyou serpentines support a more mesic, though unique, flora; the greatest number of serpentine endemics are in this province (including southwestern Oregon).

23. The unique and stressful chemical and physiological makeup of the serpentine habitat probably influences faunal adaptation and distribution, but little is known about the possible responses of native (and introduced) fauna to serpentine habitats. Specialization of insect herbivory on serpentine flora has been noted for the butterflies *Euphydryas editha* and *Pieris* spp.; serpentine species of the mustard genus *Streptanthus* apparently respond with egg mimicry to predation by *Pieris*. Herbivory as well as pollination syndromes involving serpentine biota are likely to be identified through further study.

24. Several possible explanations (hypotheses) for the origin and diversification of adaptation to serpentine have been proposed. Most are extensions of conventional ideas for the evolution of adaptedness. Plant populations may be viewed as exhibiting various stages of accommodation to serpentine, from ecotypic differentiation to unique and monotypic serpentine endemics. Some populations may be held at the early stage of ecotypic differentiation, while others may have gone through several stages to become narrow endemic species.

25. Possible modes (and/or stages) of evolutionary accommodation to serpentine include 1. ecotypic differentiation; 2. "drift" and/or depletion of biotypes; 3. catastrophic selection and saltational speciation; 4. gradual allopatric speciation with ecogeographic specialization; 5. hybridization with or without polyploidy. Only the first of these modes has been demonstrated for serpentinicolous taxa.

26. In the genus *Streptanthus* (Cruciferae) nearly all of the 16 taxa in the subgenus Euclisia are restricted in varying degrees to serpentine. A wide range of evolutionary modes—from ecotypic differentiation, races, and infraspecific taxa with fidelity to serpentine, to narrow, local to regional, endemic species of distinctive character—seems to require more than one explanation to account for such varied and extensive diversification.

27. A variety of human activities on serpentine areas in California has had its effect on the biota. Mining and geothermal power development have had the greatest effect. The

quicksilver mines of New Idria, in operation since the early 19th century, also altered the habitat for the flora by harvesting for mine timbers nearby unique stands of conifers, by building roads on the unstable serpentine, and by causing other disturbance to the remarkable flora. Geothermal power development on the serpentines of the Mayacamas range (the Geysers area of Napa, Lake, and Sonoma counties) continues to make inroads on the flora. The serpentines of northwestern California may soon be drastically altered for the extraction of nickel, especially in the Gasquet area of Del Norte County. Agriculture and timber harvest have taken place on a scale more modest than that of mining; upland serpentine soils are used marginally as range-land and some alluvial serpentines have been cropped, with much fertilizer application. It is probable that such uses have affected the viability of rare serpentine taxa.

28. Federal and state land managers recognize the limited utility of serpentine habitats for forestry and agriculture. The Soil Conservation Service consistently recommends their use primarily as habitat for wildlife and for watershed. Only limited success has been achieved in reclaiming serpentine "brush-land" by eradication of the chaparral and fertilization and seeding with palatable grasses. Serious erosion follows disturbance, further justifying the low-impact use of serpentine areas.

29. Intentional preservation of samples of outstanding serpentine habitat has been sporadic. Only one Federal Research Natural Area (Frenzel Creek in Colusa County) has been set aside to preserve a sample of the serpentine syndrome. Other state and federal land holdings do preserve serpentine barrens, but only as part of a larger preserve perimeter (e.g., state parks on Mt. Tamalpais and Mt. Diablo). U.S. National Forest and California State Forestry lands often have extensive areas of serpentine habitat; most are subject to multiple-use management. The need exists for more definitive preservation of samples of this unique ecosystem.

30. A substantial number of serpentine-endemic taxa are rare enough to be on the federal and California rare and endangered species lists. Preservation of serpentine habitat is an expected outcome of recognizing a rare plant's endangered status. Some examples include *Arabis constancei, Calochortus obispoensis, Clarkia franciscana, Eriogonum siskiyouense, Fritillaria falcata, Haplopappus ophitidis, Lilium wigginsii, Phacelia dalesiana, Streptanthus brachiatus, S. niger,* and *Veronica copelandii.*

Appendix A

Appendix A
Distribution of Serpentinite and Other Ultramafic Outcrops[1]

I. South Coast Ranges

  A. Outer Coast Ranges, Santa Barbara Co. N to San Francisco
     Bay Region

     SANTA MARIA sheet

       1. Santa Inez Mts.: Little Pine Fault area; Figueroa Mtn.,
         Santa Barbara Co.

       2. SW of Lompoc, Santa Barbara Co.

       3. SW of Guadalupe, Santa Barbara Co.

     SAN LUIS OBISPO sheet

       1. Coastal region of San Luis Obispo Co.: San Luis Range,
         Cuesta Pass area; Santa Lucia Mts.: San Rafael Mts. border
         area, San Luis Obispo/Santa Barbara cos. line

       2. Santa Lucia Mts. (western slopes): Monterey/San Luis Obispo
         cos. line

     SAN JOSE sheet (western half)

       1. E and W of Hwy 101 in vicinity of Coyote, Santa Clara Co.
         (western outcrops in Santa Cruz Mts.)

     SANTA CRUZ sheet (western half)

       1. One tiny outcrop near Pt. Sur, Monterey Co.

     SAN FRANCISCO sheet

       1. Crystal Springs Reservoir area, San Mateo Co.

       2. SE-trending band across city of San Francisco, including
         Presidio, San Francisco Co.

       3. Oakland/Berkeley Hills, Alameda/Contra Costa cos.

       4. Mt. Tamalpais and Tiburon peninsula, Marin Co.

  B. Inner south Coast Ranges

     SAN LUIS OBISPO sheet (eastern half)

       1. E of Cholame valley, in Temblor Range—e.g., at Kings/San
         Luis Obispo cos. line

       2. Stone Corral Canyon

       3. Diablo Range: Mine Mtn., Table Mtn., Kings Co.

SANTA CRUZ sheet (eastern half)

1. Southern Diablo Range: San Benito Peak, New Idria, Priest Valley; no ultramafics on W side of San Benito Valley; Monterey and San Benito cos.
2. Scattered outcrops on E slope of Diablo Range—e.g., Piedra Azul Spring, Ortigalita Peak, western Merced Co.

SAN JOSE sheet (eastern half)

1. Mt. Diablo
2. Northern Diablo Range: Mt. Hamilton area: Cedar Mts. Ridge, Red Mtn., and south along Ortigalita Fault (E slope of Diablo Range), Santa Clara and Stanislaus cos.

II. North Coast Ranges

SAN FRANCISCO sheet

1. Mt. Tamalpais and Tiburon peninsula, Marin Co.

SANTA ROSA sheet

1. Sonoma Co.: Bohemian Grove/Camp Meeker/Occidental area; The Cedars (E Austin Creek, Layton Mine); E of Santa Rosa, Buzzard Peak; Alexander Valley; Mayacamas Range
2. Napa Co.: W of Rutherford (Inglenook Ranch); Mt. St. Helena, N, W, and E slopes; W and N of Lake Berryessa
3. Lake Co.: Middletown W to Lakeport (Cobb Valley); Middletown N and E to Lake Berryessa (e.g., Butts Creek); Lower Lake/Knoxville/Lake Berryessa
4. Mendocino Co.: W of Hopland (Snow Mtn., Duncan's Peak)

UKIAH sheet

1. N/S trending massif and extensive Stony Creek fault from Wilbur Springs and Stonyford to Paskenta, traverses Colusa, Glenn, and Tehama cos.
2. Outcrops, long and narrow, from Indian Valley Ranch to Lake Pillsbury, Lake Co.
3. Eden Valley Ranch area, Mendocino Co.
4. Little Red Mtn. at and just N of Cummings (U.S. Hwy 101), Mendocino Co.

REDDING sheet
1. W of Miranda, small outcrops, Humboldt Co.
2. Mt. Lassen, eastern Humboldt and southern Trinity cos.
3. Yolla Bolly region: Little Red, Red, and Dubakella mts., borders of Trinity/Shasta/Tehama cos.; also local outcrops N to Trinity River
4. N and S of Willow Creek, Trinity River area
5. N and W of Trinity River reservoir area
6. Long, thin, NW-trending outcrops, NW and SW of Junction City, Trinity Co.
7. Twin Springs/Valentine Ridge area, SE sector of Redding sheet

WEED sheet
1. Eastern sector of sheet: Scott Mts. (Castle Crags area, granite/ultramafic contact), Mt. Eddy, upper Trinity River area, Trinity and Siskiyou cos.
2. Along Klamath River: Gottville to Horse Creek N to Oregon state line (many small and medium-sized outcrops)
3. Horse Creek to Happy Camp: N to Applegate River and S to Shackleford Creek (Salmon Mts.); Happy Camp to Weitchpec, mostly N of Salmon River canyon—e.g., Clear Creek, Cottage Grove, Somes Bar
4. Preston Peak and Sanger Peak area, N to Oregon border and S to Dillon Mtn. area
5. Western sector of sheet: from Oregon border S to Hoopa area on Trinity River—e.g., from S to N: Telescope Peak, Burrill Peak, Blue Creek Mtn., Red Mtn., Rattlesnake Mtn., Buck Mtn., Gasquet area (Gasquet Mtn.), Gordon Mtn. (?), Peridotite Canyon, High Plateau Mtn.

III. Sierra Nevada
BAKERSFIELD sheet
1. One small ultramafic outcrop just S of Hwy 499, above Kern Mesa and below Bear Mtn. (Caliente area), Kern Co.
2. Piute Mts. may have some mafic soils—e.g., Bodfish to Saddle Springs mapped as <u>bi</u> (basic intrusive), Kern Co.

FRESNO sheet

1. E of Porterville: Frazier Valley, Success Reservoir,
   Tulare Co.
2. E of Lindsay: Round Valley, Youngs Creek, and Monument
   Hill, Tulare Co.
3. E of Ivanhoe, Tulare Co.
4. Just N of Dinuba (small isolated outcrop), Fresno Co.
5. Piedra/Trimmer/Hog Mtn., large outcrop, Fresno Co.

MARIPOSA sheet

1. Buckeye Mtn. area: two small outcrops NW of Coarsegold
   (Indian Hill/French Gulch), Madera Co.
2. Extreme W central sector: W/NW of Mormon Bar and
   Mariposa, Mariposa Co.

SAN JOSE sheet (eastern half)

1. Mariposa/Bagby/Coulterville: long, narrow NW-trending
   outcrop; also extends intermittently to Jamestown,
   Mariposa Co.
2. Just W of above are massive outcrops: Red Hills,
   Chinese Camp to Copperopolis, Mariposa and Calaveras cos.

SACRAMENTO sheet

1. A more or less continuous, narrow band of ultramafics from
   Copperopolis (San Jose sheet) along Bear Mtn. fault zone N
   to Auburn (N edge of Sacramento sheet), traverses Calaveras,
   Amador, and Eldorado cos.
2. The above narrow band fans out E and W into numerous thin
   streaks of ultramafics from Cosumnes River (Big Canyon
   Creek) to N edge of sheet
3. Isolated outcrops E of the above:
   a. River Pines on Cosumnes River
   b. Ashland Creek W of Barton
   c. Placerville/Fort Jim
   d. Just NE of Esperanza Station
   e. Numerous small outcrops E of Angel's Camp

CHICO sheet

1. Melones fault zone: extending from R11E, T14N-R8E, T25N
   (from Paragon, S of Montevista, Hwy 40 N and NW to Feather

River [off map]), Placer, Nevada, Sierra, Yuba, Butte, and
Plumas cos.

2. Several outcrops from S to N of R7E, R8E—e.g., Trailer
Hill, Grass Valley, Butte/Yuba cos. line E of Forbestown

3. Extensive series of E/W outcrops either side of Feather
River Canyon from Magalia E to Grizzly Mtn. and Hartman
Bar Ridge, Butte and Plumas cos.

WESTWOOD sheet

1. From Grizzly Peak area (R10E, T25N) to S end of Lake
Almanor, Plumas Co.

2. Continuation of Melones fault zone (R7E, T25N-T26N), near
junction N and S forks Feather River; Red Hill, 6341 ft.,
Plumas Co. This terminates the northern extension of
ultramafic outcrops in the Sierra Nevada; northward, lavas
and volcanics take over)

ALTURAS sheet

No ultramafics anywhere in this vast volcanic area

IV. Southern California

LOS ANGELES sheet (continues SANTA MARIA sheet)

1. Additional serpentines in Santa Inez Mts.—e.g., Cachuma
Saddle, Santa Barbara Co.

SANTA ANA sheet

1. Basic intrusives (bi) in Pala and Gavilan Mts. of Orange
and San Diego cos.

2. Small outcrops of serpentinite in Santa Ana Mts. (Vogl 1973)

3. Small outcrop in extreme NE section, Hexie Mts., N of
Little San Bernardino Mts., San Bernardino Co.

[1]Based on Geologic Map of California (Calif. Div. of Mines and Geology,
1958-1967).

Appendix B

Soil Series Derived from Serpentinite Rock, distribution by county or area[1]

County or area publ. date/acreage	Soil series	Acreage	Percentage of survey area	General location	Use and/or remarks
Alameda area/1966 325,000 acres	HENNEKE	9,044 acres	2.8%	Vicinity of Cedar Mtn.	WA, WI, RE*
Amador area/1965 298,992 acres	HENNEKE	2,064	0.7	N from Mokelumne River to El Dorado Co. line	Grazing, WA, WI, RE
	Serpentine rock land	3,640	1.2		
Colusa area/1907 2,500,000 acres	Rough stony land	3,392	0.7	Foothills W of Maxwell	Probably includes serpentine soils
El Dorado area/1974 539,065 acres	DELPIEDRA	1,592	0.3	NW sector, vicinity of Greenwood	Range and WA
	Serpentine rock land	18,201	3.4	Near Folsom Reservoir	
Eastern Fresno/1971 1,109,156 acres	FANCHER	9,444	0.4	Foothills N of Kings River	Range
	DELPIEDRA	5,576	0.5	Foothills N of Kings River	Grazing

County/Year/Acres	Soil series	Acres	%	Location	Notes
Glenn Co./1968 846,080 acres	DUBAKELLA	52	0.1	W (upland) sector of county	WA, WI
	HENNEKE	11,945	1.4		
	MONTARA	137	0.1		
	POLEBAR	2,140	0.2		
Mariposa Co. area/1974 931,200 acres	HENNEKE	4,942	1.0	Bagby area, NE sector	Limited range WA, WI
Napa Co./1978 485,120 acres	HENNEKE	56,815	4.4	Near Butts Canyon, S of Snell Valley, around Cedar Valley and Adams Ridge W of Knoxville/Lake Berryessa Road	Woodland and chaparral, 500-4000 ft. elev., WA, WI, RE and some grazing
	MONTARA	7,690	1.6		
Monterey Co./1978 2,127,360 acres	HENNEKE	7,753	0.3	Big Sur area; scattered outcrops in Santa Lucia and Diablo ranges	Range, WI, WA, RE
	MONTARA	11,115	0.5		
	CLIMARA	3,260	0.2		
Nevada Co. area/1975 341,966 acres	DUBAKELLA	1,325	0.1	NNW of Grass Valley	WA, WI, grazing; 2220-2700 ft. elev.
Placer Co., western part/1980 411,544 acres	DUBAKELLA	1,950	0.5	N and E of Auburn	

Location / Year / Acres	Soil type	Acres	Percent	Locality	Notes
	HENNEKE	770	0.2		
	Rock land	7,500	1.8		A portion is serpentinite
San Benito Co./1969 893,440 acres	HENNEKE	19,980	2.2	S sector of county	WA, WI, RE
	Igneous rock land	30,650	3.4		
	MONTARA	2,490	0.3		
	CLIMARA	6,150	0.7		
San Francisco Bay Region/1914 2,517,120 acres	Rough stony land	28,929	1.1	Many upland areas bordering alluvial valleys	Includes serpentinite areas
	Stony soils, un-differentiated	123,264			
San Luis Obispo area/1928 469,760 acres	Rough, broken and stony land	10,112	2.1	Various localities	May include serpentine soils
San Mateo area/1961 168,898 acres	MONTARA	133	0.1	NE part at 2000 ft. elev.	WA
Northern Santa Barbara Co./1972 830,870 acres	MONTARA	908	0.1	Vicinity of Figueroa Mtn. Mtn.	Range, WA, WI

Survey area/year	Soil	Acres	%	Location	Remarks
Santa Clara area/1958 314,000 acres	MONTARA	14,024	4.5	Hills in S sector	Grazing
Eastern Santa Clara area/1974 519,280 acres	HENNEKE	2,650	0.5	Red Mtn. area bordering Stanislaus Co.	WA, WI, RE
	MONTARA	6,360	1.2		
Shasta Co. area/1974 1,035,000 acres	HENNEKE	1,325	0.1	Uplands W of Ono	Range, WA, WI; 1000-2500 ft. elev.
Solano Co./1977 526,720 acres	CONEJO	ca 1,400	0.3		Basic igneous alluvium (non-serpentine)
Sonoma Co./1972 1,010,560 acres	HENNEKE	14,189	1.4	NE half of county (Geysers/Mayacamas Mts.); also in N sector (Austin Creek drainage) S to Occidental	WA, WI, RE, geothermal
Tehama Co./1967 1,904,640 acres	DUBAKELLA	3,650	0.1	W sector	WA, WI
	HENNEKE	24,713	1.3	Mountainous W sector	WA, WI
Ukiah area/1914 193,920 acres	Rough mountainous land	131,648	67.1	Hill land bordering valleys	May include serpentine soils

Yolo Co./1972 661,760 acres	Rock land	36,139	5.5	W border	Includes serpentinite; WA, WI
	CLIMARA	324	<0.1	W border	

[1]Compiled from relevant Soil Surveys, USDA Soil Conservation Service, on file at the University of Washington Library. Dubakella, Henneke, Ipish, Montara, Obispo, Wadesprings, and Weitchpec are recently established California series in the serpentinitic families. These series are in several soil groups: Haploxerolls, Argixerolls, and Xerochrepts (E.B. Alexander, U.S. Forest Service, personal communication).

*WA = watershed; WI = wildlife; RE = recreation.

Appendix C

Taxa endemic to ultramafic substrates in California

Taxon	Fidelity to JUB[1]	Distribution	Sample localities[2]
FERNS			
Aspidotis carlotta-hallii (Wagner & Gilbert) Lellinger	+++	San Luis Obispo to Marin cos. (Coast Ranges)	Marin Co.: Rifle Camp, Mt. Tamalpais State Park
Polystichum lemmonii Underw.	+++	Trinity, Del Norte, and Siskiyou cos. N to Washington	Siskiyou Co.: slopes of Mt. Eddy
CONIFERS			
Cupressus macnabiana A. Murr. (may be indicator, only)	+++?	Sierra Nevada from Amador to Butte cos.; Coast Ranges from Shasta to Sonoma cos.	Butte Co.: Magalia area
C. sargentii Jeps.	+++	Santa Barbara to Mendocino cos. (Coast Ranges)	Santa Barbara Co.: Figueroa Mtn. area; Napa Co.: St. Helena Creek area
Juniperus communis L. var. jackii	+++	Del Norte Co. to SW Oregon	Del Norte Co.: above Eighteen Mile Creek on old Gasquet Toll Road
MONOCOTYLEDONS			
Allium falcifolium H. & A.	++	Santa Cruz Co. to SW Oregon	San Mateo Co.: Jasper Ridge
A. fimbriatum Wats. var. diabloense Ownbey	+++	Stanislaus to Santa Barbara cos.	San Benito Co.: Clear Creek, New Idria area
A. fimbriatum Wats. var. purdyi (Eastw.) Ownbey & Aase	+++	Lake/Colusa cos. line	On road to Williams

A. fimbriatum Wats. var. *sharsmithiae* Ownbey & Aase	+++	Stanislaus Co.	Near head of Arroyo del Puerto
A. hoffmanii Ownbey	+++	Shasta, Tehama, and Trinity cos.	Trinity Co.: Black Lassic
A. howellii Eastw. var. *sanbenitensis* (Traub) Ownbey & Aase	++	San Benito Co.	E of San Carlos Peak, New Idria
A. sanbornii Wood	++	Sierra Nevada: Shasta to Calaveras cos.	Mariposa Co.: 3 mi. NW of Coulterville
Brodiaea stellaris Wats.	+++?	Humboldt to Sonoma cos.	Sonoma Co.: Harrison Grade
Calamagrostis ophitidis (J.T. Howell) Nygren	+++	Marin to Sonoma and Lake cos.	Marin Co.: J.T. Howell Botanical Garden, Tiburon peninsula
Calochortus clavatus Wats.	++	Sierra Nevada: Eldorado to Mariposa cos.; Coast Ranges: Stanislaus to Los Angeles cos.	(None seen)
C. coeruleus var. *fimbriatus* Ownbey	++	Siskiyou to Lake cos.	(None seen)
C. greenei Wats.	+++?	Siskiyou Co.; Jackson Co., Ore.	(None seen)
C. obispoensis Lemmon	+++	San Luis Obispo Co.	2.4 mi. N of Cuesta Pass
C. tiburonensis A.J. Hill	+++	Marin Co.	Tiburon peninsula
C. vestae Purdy	++?	North Coast Ranges, Humboldt to Napa and Sonoma cos.	Humboldt Co.: Dobbyn Creek, 3 mi. NE of Alderpoint
Carex obispoensis Stacey	+++	San Luis Obispo Co.	Steiner Creek near San Luis Obispo

Species	Rating	Distribution	Locality
Chlorogalum grandiflorum Hoov.	+++	Tuolumne Co.	3 mi. N of Keystone
C. pomeridianum (DC) Kunth var. *minus* Hoov.	+++	Tehama and Glenn cos.	Glenn Co.: Alder Springs road 20 mi. W of Willows
Erythronium californicum Purdy	++?	Humboldt and Shasta to Sonoma and Colusa cos.	Humboldt Co.: Horse Mtn.
E. citrinum Wats.	++	Siskiyou and Del Norte cos. to SW Oregon	Del Norte Co.: 3 miles E of Gasquet
E. helenae Applegate	+++	Mt. St. Helena area (borders of Napa, Sonoma, and Lake cos.)	Lake Co.: steep walls of side stream, Bulls Canyon
E. hendersonii Wats.	++?	Siskiyou Co. and SW Oregon	Siskiyou Co.: Klamath River between Hamburg and Seiad Valley
E. howellii Wats.	++	Del Norte Co. to SW Oregon	Del Norte Co.: Happy Camp road, 9 mi. SE of Takilma
E. multiscapoideum (Kell.) Nels. and Kenn.	+++?	Tehama and Butte to Mariposa cos.	Butte Co.: Magalia area
E. tuolumnense Applegate	++?	Tuolumne and Stanislaus cos.	Tuolumne Co.: $\frac{3}{4}$ mi. SW of Italian Bar, S fork Stanislaus River
Fritillaria falcata (Jeps.) D.E. Beetle	+++	Stanislaus, Santa Clara, and San Benito cos.	Stanislaus Co.: Arroyo del Puerto
F. glauca Greene	++	Lake, Tehama, and Humboldt cos. to SW Oregon	(California specimer not seen) Josephine Co., Ore.: mtn. slopes near O'Brien
F. purdyi Eastw.	+++	Humboldt and Trinity to Napa cos.	Humboldt Co.: Horse Mtn.

Species		Distribution	Locality
F. recurva Benth. var. *coccinea* Greene	+++	Mendocino to Napa cos.	Napa Co.: E slope of Mt. St. Helena
Lilium bolanderi Wats.	+++	Del Norte Co. to SW Oregon	Del Norte Co.: 1 - 2 mi. SE of Elk Valley, 2 mi. SE of Chimney Rock
L. kelloggii Purdy	++?	Del Norte and Humboldt cos.	Humboldt Co.: on road to Bald Mtn., E of Korbel
Odontostomum hartwegii Torr.	++?	Sierra Nevada: Butte to Mariposa cos.; Coast Ranges: Tehama and Napa cos.	El Dorado Co.: Serpentine Plateau between Green Valley and Eggars
Poa piperi Hitchc.	++?	Del Norte Co., S Oregon	Near Gasquet, N side of Smith River
Stipa lemmonii (Vasey) Scribn. var. *pubescens* Crampt.	+++	Tehama Co. N to Oregon border	Trinity Co.: 100 yds. N of Hidden Lake; Underwood Mtn., SW of Burnt Ranch

DICOTYLEDONS

Species		Distribution	Locality
Acanthomintha lanceolata Curran	+++	Alameda to San Benito cos.	San Benito Co.: Panoche Pass, ca. 6 mi. SE of Paicines, San Carlo Range
A. obovata Jeps. subsp. *duttonii* Abrams	+++	San Mateo Co.	Crystal Springs Lake
Antennaria suffrutescens Greene	+++	Del Norte and Siskiyou cos. to SW Oregon	Humboldt Co.: summit area, Horse Mtn.
Arabis aculeolata Greene	+++	Del Norte Co. to SW Oregon	(No Calif. specimens seen) Josephine Co., Oregon: Rogue River, 6 mi. SE of Galice

A. constancei Rollins	+++	Plumas Co.	2 mi. from Spring Garden on road to Quincy
A. mcdonaldiana Eastw.	+++	Mendocino Co.	Summit "Big" Red Mtn., 5 mi. SE of Bell Springs
A. serpentinicola Rollins	+++	Siskiyou Co.	Preston Peak
Arctostaphylos hookeri G. Don ssp. *franciscana* (Eastw.) Munz	+++	San Francisco	Laurel Hill Cemetery
A. obispoensis Eastw.	+++	San Luis Obispo and S Monterey cos.	San Luis Obispo Co.: on ridge 4 mi. SE of Cuesta summit on Hwy 101
A. stanfordiana Parry ssp. *hispidula* (Howell) Adams	+++	Del Norte Co.	Low Divide
A. stanfordiana Parry var. *bakeri* (Eastw.) Adams	+++	Sonoma Co.	Harrison Grade between Occidental and Camp Meeker
A. viscida Parry	++?	Sierra Nevada: Kern Co.; N Coast Ranges: Lake and Napa cos.	(None examined from serpentine; possibly only an indicator species)
Arenaria douglasii Fenzl. ex T.&G. var. *emarginata* H.K. Sharsm.	+++	Santa Clara Co.	Red Mts., Mt. Hamilton Range
A howellii Wats.	+++	Del Norte and Trinity cos.	Del Norte Co.: 7 mi. E of Smith River Village
A. rosei Maguire & Barneby	+++	Trinity and Tehama cos.	Trinity Co.: at Peanut
Arnica cernua Howell	+++	Humboldt, Del Norte, and Siskiyou cos. to SW Oregon	(No Calif. specimens seen)

Species		Distribution	Locality
Asclepias solanoana Woodson	+++	Lake to Trinity cos.	Trinity Co.: headwaters of Salt Creek, ca. 8 mi. W of Idyllwild
Astragalus breweri Gray	+++	Mendocino and Lake to Marin cos.	Lake Co.: 2 mi. N of Middletown
A. rattanii Gray var. *jepsonianus* Barneby	+++	Lake, Napa, Colusa, and Tehama cos.	Napa Co.: $\frac{1}{2}$ mi. NW of Knoxville
A. whitneyi Gray var. *siskiyouensis* (Rydb.) Barneby	+++	Trinity to Siskiyou cos. and SW Oregon	Trinity Co.: 1 mi. SW of Peanut
Balsamorhiza macrolepis Sharp	++?	San Francisco Bay Region; Sierra Nevada: Mariposa to Butte cos.	Tuolumne Co.: near Big Meadows, Coulterville road
Benitoa occidentalis (Hall.) Keck (= *Haplopappus occidentalis* Hall)	++	San Benito, Fresno, and Monterey cos.	Fresno Co.: N side of Parkfield Grade
Calycadenia pauciflora Gray	+++	Sierra Nevada: Calaveras and Stanislaus cos.; Coast Ranges: Glenn to Sonoma cos.	Lake Co.: 2 mi. NE of Middletown
Camissonia benitensis Raven	+++	San Benito Co.	San Benito Co.: New Idria area
Castilleja brevilobata Piper	+++	Del Norte Co. and adjacent Josephine Co., Ore.	Del Norte Co.: along road to Monumental, 1 mi. NE of Gasquet
C. miniata Hook. ssp. *elata* (Piper) Munz (= *C. elata* Piper, fide L. Heckard)	+++	Del Norte and Siskiyou cos.	Del Norte Co.: Smith River and Myrtle Creek
C. neglecta Zeile in Jeps.	+++	Marin Co.	Tiburon
Ceanothus ferrisae McMinn	+++	Santa Clara and Santa Cruz cos.	Santa Clara Co.: 3 mi. E of Madrone

C. jepsonii Greene	+++	Marin to Mendocino cos.	Sonoma Co.: Harrison Grade ca. 2 mi. NNE of Occidental
C. jepsonii var. *albiflorus* J.T. Howell	+++	Napa and Lake cos.	Lake Co.: 4 mi. NE of Middletown (likely site)
C. pumilus Greene	+++	Mendocino and Trinity to Del Norte and Siskiyou cos.; SW Oregon	Del Norte Co.: French Hill, 2 mi. S of Gasquet
Chorizanthe breweri Wats.	++?	San Luis Obispo Co.	1.3 mi. N of San Luis Obispo
C. uniaristata T.&G.	++?	San Benito to Santa Barbara and Kern cos.	Fresno Co.: Little Panoche Creek, $\frac{1}{2}$ mi. NE of old school
Cirsium breweri (Gray) Jeps. (possibly on serpentine only in Coast Ranges)	++	Monterey to Del Norte and SW Oregon E to W Nevada	Humboldt Co.: Ruby Creek on the Willow Creek Hwy
C. campylon H.K. Sharsm.	++	Stanislaus and Santa Clara cos.	Santa Clara Co.: side canyon by Metcalf Road near Coyote
C. fontinale (Greene) Jeps.	+++	San Mateo Co.	Near Crystal Springs Lake
C. fontinale (Greene) Jeps. var. *obispoense* J.T. Howell	+++	San Luis Obispo Co.	Chorro Creek, $\frac{2}{10}$ mi. below junction forks of San Simeon Creek
C. vaseyi (Gray) Jeps.	+++	Marin Co.	Mt. Tamalpais, in the Portrero
Clarkia franciscana Lewis & Raven	+++	San Francisco	The Presidio
Collinsia greenei Gray	+++	Sonoma to Trinity cos.	Sonoma Co.: Pine Flat/ Middletown road, Mayacamas Mts.

Species		Distribution	Locality
Collomia diversifolia Greene	+++	Napa to Mendocino cos.	Lake Co.: 3 mi. E of Middletown
Convolvulus malacophyllus Greene subsp. *collinus* (Greene) Abrams	+++	Mendocino to San Benito cos.	Sonoma Co.: summit of grade between Mark West Valley and Franz Valley
Cordylanthus nidularius J.T. Howell	+++	Contra Costa Co.	Near Deer Flat, Mt. Diablo
C. pringlei Gray	+++	Lake Co.	3.5 mi. E of Middletown
C. tenuis Gray subsp. *brunneus* (Jeps.) Munz	+++	Napa, Sonoma, and Lake cos.	Napa Co.: along St. Helena Creek, just S of county line
C. tenuis ssp. *capillaris*	+++	Sonoma Co.	Sonoma Co.: 2 mi. E of Occidental
Cryptantha clevelandii Greene var. *dissita* (Jtn.) Jeps. & Hoov.	+++	"Serpentine outcrops, near Lakeport, Lake Co." (Munz and Keck, 1959, p. 574)	(None seen)
C. hispidula Greene ex Brand	+++	Napa and Lake cos.	Lake Co.: "Hill 1030," ca. 3 mi. N of Middletown
C. mariposae Jtn.	+++	Calaveras to Mariposa cos.	Mariposa Co.: grade from Coulterville to Bagby
Delphinium parryi Gray var. *eastwoodiae* Ewan	+++	San Luis Obispo Co.	1st ridge W of Cerro Romauldo
D. uliginosum Curran	+++	Napa, Lake, and Colusa cos.	Napa Co.: 3 mi. W of Knoxville
Dentaria gemmata (Greene) Howell	++?	Del Norte Co. to SW Oregon	(No Calif. specimen seen) Josephine Co., Ore.: near top of Tennessee Pass near Kirby

Species		Distribution	Locality
D. pachystigma Wats. var. *dissectifolia*	+++?	Butte and Mendocino cos.	Butte Co.: near Magalia
Dicentra formosa (Andr.) Walp. ssp. *oregona* (Eastw.) Munz	+++	Del Norte Co. and SW Oregon	(No Calif. specimens seen) Josephine Co., Ore.: 8 mi. W of O'Brien
Dudleya abramsii Rose ssp. *murina* (Eastw.) Moran	+++	San Luis Obispo Co.	Along Hwy 101 just E of San Luis Obispo
D. bettinae Hoov.	+++	San Luis Obispo Co.	1 mi. S of Cayucos and 1 mi. W of Cerro Romauldo
Emmenanthe penduliflora Benth. var. *rosea* Brand	+++	Santa Clara to Ventura cos.	San Benito Co.: Vasquez Creek of Little Panoche Valley
Epilobium rigidum Hausskn.	+++	Del Norte Co. to SW Oregon	Del Norte Co.: 18-mile Creek, Smith River
Eriogonum alpinum Engelm.	+++	Siskiyou Co.	Near summit of Mt. Eddy
E. argillosum J.T. Howell	+++	Santa Clara, San Benito, and Monterey cos.	(No specimens seen)
E. caninum (Greene) Munz	++	Marin and Contra Costa cos.	Marin Co.: Tiburon
C. congdonii (Stokes) Reveal	+++?	NW Calif.	Trinity Co.: along Calif. Hwy 3, 0.9 mi. S of Scott Mtn. pass
E. covilleanum Eastw.	++	Alameda to Kern cos.	San Benito Co.: Panoche road 3 mi. below Idria
E. kelloggii Gray	+++	Mendocino Co.	Red Mtn.

Species		Distribution	Locality
E. libertini Reveal	+++	Shasta, Tehama, and Trinity cos.	Tehama Co.: $\frac{3}{4}$ mi. N of Tedoc Gap
E. siskiyouense Small	++	Siskiyou and Trinity cos.	Siskiyou Co.: Mt. Eddy
E. ternatum Howell	+++	Del Norte and Siskiyou cos. to SW Oregon	Del Norte Co.: 1 - 2 mi. S of Elk Valley, at head of Blue Creek, 1 mi. E of Chimney Rock
E. tripodum Greene	+++	Tehama and Lake cos. of Coast Ranges; Mariposa and Tuolumne cos. of Sierra Nevada	Tuolumne Co.: 3 mi. N of Keystone; Colusa Co.: between Ladoga and Cooke Springs
Eriophyllum jepsonii Greene	++	Contra Costa to San Benito cos.	Stanislaus Co.: near head of Arroyo del Puerto, Red Mts.
E. latilobum Rydb.	+++	San Mateo Co.	Crystal Springs Lake region
Fremontodendron decumbens R. Lloyd	++?	El Dorado Co.	Ridge S of Pine Hill (on olivine gabbro)
Galium ambiguum var. *siskiyouense* Ferris	+++	Trinity and Humboldt cos. to Siskiyou Mts.	Del Norte Co.: French Hill, 2 mi. S of Gasquet
G. andrewsii Gray var. *gatense* Dempster	++	Fresno and San Benito cos.	San Benito Co.: Clear Creek, New Idria area
G. hardhamiae Dempster	+++	San Luis Obispo Co.	Head of Arroyo del la Cruz and Little Burnett creeks, Santa Lucia Mts.
Garrya congdonii Eastw.	+++	Sierra Nevada: Mariposa Co. N; Coast Ranges: Tehama to San Benito cos.	San Benito Co.: 7 mi. from Hernandez on road to New Idria

Gentiana bisetaea Howell	+++	SW Oregon; possibly Del Norte Co.	Josephine Co., Ore.: Bogs at base of Oregon Mtn., among *Darlingtonia*
G. setigera Gray	++	Mendocino Co. to SW Oregon	Mendocino Co.: Red Mtn.
Githopsis pulchella Vatke	+++	Amador to Mariposa cos.	Tuolumne Co.: 2 mi. SW of Chinese Camp, Red Hills
Haplopappus ophitidis (J.T. Howell) Keck	+++	NW Tehama Co.	Summit of Mt. Tedoc, 5000 ft.; 2 mi. NE of White Rock Ranger Station
H. racemosus (Nutt.) Torr. subsp. *congestus* (Greene) Hall	+++	Del Norte Co. to SW Oregon	Del Norte Co.: French Hill, Gasquet
Helianthus bolanderi Gray subsp. *exilis* (Gray) Heiser (= *H. exilis* Gray)	+++	Coast Ranges: San Luis Obispo to Trinity cos.; Sierra Nevada: Tuolumne to Plumas cos.; SW Oregon	Napa Co.: 3.9 mi. NW of Knoxville on road to Lower Lake
Hesperolinon adenophyllum (Gray) Small	+++	Mendocino and Lake cos.	Lake Co.: at saddle between Elk Mtn. and Pine Mtn.
H. bicarpellatum (Shars.) Shars.	+++	Napa and Lake cos.	Napa Co.: between Walter Springs and Samuel Springs above Pope Creek
H. congestum (Gray) Small	+++	Marin to San Mateo cos.	Marin Co.: Tiburon Peninsula
H. didymocarpum H.K. Sharsm.	+++	Lake Co.	W of Big Canyon Creek
H. disjunctum H.K. Sharsm.	+++	Tehama to Fresno and Monterey cos.	Colusa Co.: road from Stonyford to Fouts Springs
H. drymarioides (Curran) Small	+++	Colusa, Glenn, and Lake cos.	Glenn Co.: Black Diamond Ridge

Species		Distribution	Location
H. spergulinum (Gray) Small	+++	Mendocino, Lake, Sonoma, and Napa cos.	Sonoma Co.: near Occidental
H. tehamense H.K. Sharsm.	+++	Glenn and Tehama cos.	Tehama Co.: Paskenta grade of Covelo/Paskenta road
Hieracium bolanderi Gray	+++	Mendocino, Trinity, Del Norte, and Siskiyou cos. to SW Oregon	Del Norte Co.: Stony Creek bog near Gasquet
Horkelia sericata Wats.	+++?	Humboldt Co. to SW Oregon	Del Norte Co.: near top of Coon Mtn., 3900 ft.
Lagophylla minor (Keck) Keck	+++?	Napa to Glenn cos. in Coast Ranges; El Dorado Co. in Sierra Nevada	Napa Co.: 3 mi. W of Knoxville on road to Lower Lake
Layia discoidea Keck	+++	San Benito Co.	Along Clear Creek, 8.7 mi. S of New Idria
Lessingia ramulosa Gray var. *glabrata* Keck	+++?	Santa Clara Co.	(None seen)
L. ramulosa Gray var. *micradenia* (Greene) J.T. Howell	+++?	Marin Co.	4 mi. W of Fairfax, above Alpine Lake
L. ramulosa Gray var. *ramulosa*	+++?	North Coast Ranges	Lake Co.: Pope Valley/ Middletown Road, 8 mi. ESE of Middletown
Lewisia stebbinsii Gankin.	+++	Mendocino Co.	Ridge of Mt. Hull
Linanthus ambiguus (Rattan) Greene	+++	Santa Clara to San Benito cos.	San Benito Co.: above New Idria Mine road, ca. 0.5 mi. W of Idria

Species		Distribution	Locality
L. dichotomus Benth. ssp. *meridianus* (Eastw.) Mason	+++	Napa to Butte cos.; El Dorado Co.	Lake Co.: end of Walker Ridge road in Kowalsil Ranch; El Dorado Co.: Garden Valley
L. liniflorus (Benth.) Greene	++	Contra Costa to San Luis Obispo cos.	(None seen)
Lithocarpus densiflora (H.&A.) Rehder var. *echincides* (R. Br.) Abrams	+++	Del Norte and Mariposa cos.	Del Norte Co.: below Camp 6 on road from Gordon Mtn. toward Bear Mtn. (some Sierra Nevada sites not serpentine?)
Lomatium ciliolatum Jeps.	+++	Glenn and Lake to Trinity and Mendocino cos.	Trinity Co.: saddle between North Yolla Bolly and Black Rock Mts.
L. ciliolatum var. *hooveri* Math. & Const.	+++	Napa, Colusa, and Lake cos.	Lake Co.: Bartlett Springs road, 4.5 mi. SW of Leesville
L. congdonii Coult. & Rose	+++	Mariposa and Tuolumne cos.	Tuolumne Co.: 3 mi. S of Chinese Camp on dirt road to La Grange
L. engelmannii Math.	+++	Trinity and Siskiyou cos. to SW Oregon	Trinity Co.: between Coffee and Eagle creeks, 6 mi. N of Carrville
L. howellii (Wats.) Jeps.	+++	Del Norte Co. to SW Oregon	Del Norte Co.: Low Divide, 7 mi. E of Smith River village
L. marginatum (Benth.) Coult.& Rose var. *marginatum*	+++	Sierra Nevada foothills from Tulare Co. N	Mariposa Co.: 5 mi. N of Coulterville
L. marginatum var. *purpureum* Jeps.	+++	Napa to Shasta cos.	Napa Co.: on hills at head of Moore's Creek, 5 mi. E of Angwin's

Species	Rating	Distribution	Locality
L. tracyi Math. & Const.	+++	Trinity and Humboldt cos.; SW Oregon	Trinity Co.: 1.5 mi. above Peanut on State Hwy 36
Lupinus lapidicola Heller	+++	Siskiyou Co.	Mt. Eddy
L. spectabilis Hoov.	+++	Mariposa and Tuolumne cos.	Mariposa Co.: 4 mi. N of Bagby
Madia hallii Keck	+++	Napa and Lake cos.	Lake Co.: 3 mi. NE of Middletown
Mimulus nudatus Greene	+++	Lake Co.	Wet crevices at Lake/Colusa cos. line
Monardella benitensis Hardham	+++	San Benito Co.	Clear Creek near New Idria
M. palmeri Gray	+++	Monterey to San Luis Obispo cos.	San Luis Obispo Co.: Rinconada Mine, between Santa Margarita and Pozo
M. villosa Benth. ssp. *neglecta* (Greene) Epl. (= *M. purpurea* Howell?)	+++	Marin and Sonoma cos. N to Del Norte Co.	Del Norte Co.: at intersection of Haynes Flat road on Coon Mtn.
M. villosa ssp. *sheltonii* (Torr.) Epl.	+++	Coast Ranges: Monterey Co. N to SW Oregon; Sierra Nevada: Tulare Co. N	Butte Co.: 23 mi. from Oroville E of Jarboe Pass, on Feather River Hwy
M. viridis Jeps.	+++	Napa and Lake cos.	Lake Co.: 0.4 mi. SE of Black Oak Villa in Butts Canyon
Montia gypsophiloides (F.&M.) Howell	++?	San Luis Obispo to Mendocino cos.	Marin Co.: above Alpine Lake, 5.4 mi. SE of Fairfax
M. spathulata (Dougl.) Howell var. *rosulata* (Eastw.) J.T. Howell	+++	Marin Co.	Rock Spring, Mt. Tamalpais

Species		Distribution	Locality
Navarretia heterodoxa (Greene) Greene ssp. *rosulata* (Brand) Mason	+++	Marin Co.	Near Liberty Camp
N. *jepsonii* V. Bailey	+++	S Lake and W Colusa cos.	Colusa Co.: Black Diamond Ridge, Mendocino Nat'l Forest
N. *mitracarpa* Greene	+++	Occasional from S Oregon through inner Coast Ranges; most abundant in Monterey and San Luis Obispo cos.	Santa Barbara Co.: serp. ridge between Camuesa Creek and Oso Creek, S of Little Pine Mtn.
N. *mitracarpa* subsp. *jaredii* (Eastw.) Mason	+++	San Luis Obispo Co. and adjacent areas	Fresno Co.: 12mi. from Coalinga on road to Parkfield
Nemacladus montanus Greene	+++	Napa and Lake cos.	Napa Co.: slopes above Soda Creek along State Hwy 28, 8.2 mi. E of Conn Dam
Parvisedum pentandrum (H.K. Sharsm.) Clausen	++?	San Benito to Lake cos.	Lake Co.: 4.4 mi. S of Lakeport
Perideridia leptocarpa Chuang & Const.	+++	Siskiyou Co.	3 mi. W of Cecilville
Phacelia breweri Gray	+++	Contra Costa to San Benito cos.	Stanislaus Co.: Arroyo del Puerto, Mt. Hamilton Range
P. *corymbosa* Jeps.	+++	Sonoma Co. to SW Oregon	Del Norte Co.: Hayne's Flat road on Coon Mtn.
P. *dalesiana* J.T. Howell	+++	Trinity and Siskiyou cos.	Siskiyou Co.: summit of Scott Mts.
P. *greenei* J.T. Howell	+++	Siskiyou Co.	Scott Mtn.

Species		Distribution	Locality
P. phacelioides (Benth.) Brand	++?	Contra Costa to Santa Clara cos.	Contra Costa Co.: Mt. Diablo, Juniper Camp
Polygonum spergulariaeforme Meissn.	++?	Lake and Nevada cos. N to British Columbia	Trinity Co.: SW of Peanut
Quercus durata Jeps.	+++	San Luis Obispo to Trinity cos. in Coast Ranges; El Dorado to Nevada cos. in Sierra Nevada	Lake Co.: along Morgan Valley road, 1 mi. W of Napa Co. line
Raillardella pringlei Greene	+++	Siskiyou and Trinity cos.	Trinity Co.: Landers Creek above Sunrise Creek
Rhamnus californica Esch. spp. *crassifolia* (Jeps.) C.B. Wolf	+++	Napa and Lake to Trinity cos.	Trinity Co.: 1 mi. SW of Peanut
Rudbeckia californica Gray var. *glauca* Blake	+++	Del Norte Co. to S Oregon	Del Norte Co.: Smith River, 3 mi. E of Gasquet
Salix breweri Bebb.	+++	San Benito to Lake and Colusa cos.	San Benito Co.: Clear Creek, 4 mi. from confluence with San Benito River, San Carlos Range
S. delnortensis C.K. Schneid	+++	Del Norte Co.	Near Smith River below Gasquet
Sanicula peckiana Macbr.	+++	Del Norte Co. to SW Oregon	Del Norte Co.: French Hill 2 mi. S of Gasquet
S. tracyi Shan & Const.	+++	Trinity and Humboldt cos. and SW Oregon	Trinity Co.: oak woods near Van Duzen River about 6 mi. SE of Cobbs on road to Ruth
Sedum albomarginatum R.T. Clausen	+++	Butte and Plumas cos.	Plumas Co.: along N fork of Feather River between Rich Gulch and Rich Bar

S. laxum (Britton) Berger ssp. *eastwoodiae* (Britton) R.T. Clausen	Glenn, Lake, and Mendocino cos.	+++	Mendocino Co.: Red Mtn., ca. 4 mi. from Bell Springs Road
Senecio clevelandii Greene	Napa and Lake cos.	+++	Lake Co.: Butts Creek canyon
S. clevelandii Greene var. *heterophyllus* Hoov.	Tuolumne Co.	+++	Tuolumne Co.: 3 mi. S of Chinese Camp in wet rocky places
S. greenei Gray	Napa and Lake to Trinity cos.	+++	Trinity Co.: Peanut
S. lewisrosei J.T. Howell (= variant of *S. eurycephalus* T.& G., in Munz and Keck)	Feather River region, Butte Co.	+++?	Near Magalia
S. ligulifolius Greene	Siskiyou and Del Norte cos.	+++?	Del Norte Co.: Patrick Creek, 1.1 mi. below summit of old road over Oregon Mtn.
Sidalcea diploscypha (T.& G.) Gray	Coast Ranges from San Luis Obispo to Humboldt cos.; Sierra Nevada foothills	+++?	Napa Co.: Pope Canyon road, 6.2 mi. W of Knoxville road
S. hickmanii Greene ssp. *anomala* C.L. Hitch.	San Luis Obispo Co.	+++	2.25 mi. NW of Cuesta
S. hickmanii ssp. *viridis* C.L.Hitch.	Marin Co.	+++	Big Carson Ridge, in burnt area
Silene campanulata Wats. ssp. *campanulata*	Mendocino Co.	+++	Red Mtn.
S. campanulata ssp. *glandulosa* Hitch. & Maguire	Lake and Glenn to Shasta, Siskiyou, and Humboldt cos.; S Oregon	+++	Trinity Co.: Lassics

Taxon		Distribution	Locality
S. hookeri Nutt. ex T.&G. ssp. *bolanderi* (Gray) Abrams	+++	Humboldt, Mendocino, and Trinity cos.	Trinity Co.: road to Peanut, 3 mi. beyond Wildwood
Streptanthus amplexicaulis (Wats.) Jeps. var. *barbarae* J.T. Howell	+++	Santa Barbara (and Ventura?) Co.	Santa Barbara Co.: Cachuma Saddle, below Figueroa Mtn.
S. barbatus Wats.	+++	Trinity Co.	Just above Salt Creek, 2 mi. SW of Peanut on Hwy 3
S. barbiger Greene	+++	Napa to Mendocino cos.	Lake Co.: Hills above Cobb Valley at Glenwood
S. batrachopus Morrison	+++	Marin (and Sonoma?) Co.	Marin Co.: Big Carson Ridge, above San Geronimo
S. brachiatus Hoffman	+++	Sonoma and Lake cos.	Sonoma Co.: near Contact Mine E of Pine Flat
S. breweri Gray	+++	San Benito to Glenn cos.	Napa Co.: roadcut, just W of Lake Berryessa on road to Pope Valley
S. drepanoides Kruckeberg & Morrison	+++	Tehama to Trinity cos.	Tehama Co.: 12 mi. W of Paskenta on Paskenta/Covelo road
S. glandulosus Hook., ssp. *pulchellus* (Greene) Kruckeberg	+++	Marin Co.	Bootjack Camp, Mt. Tamalpais State Park
S. hesperidis Jepson	+++	Napa and Lake cos.	Lake Co.: 4 mi. NE of Middletown, on Hwy 29
S. howellii Wats.	+++	Siskiyou and Del Norte cos. to SW Oregon	Del Norte Co.: Smith River drainage (No Calif. specimens seen)

Species		Distribution	Locality
S. insignis Jeps. var. *insignis*	+++	Fresno, San Benito, and Monterey cos.	San Benito Co.: 6 mi. W of Panoche Pass
S. insignis var. *lyonii* Kruckeberg & Morrison	++	Merced Co. (western)	Above Wisenor Flats on Arburua Ranch
S. morrisonii Hoffman	+++	Sonoma, Napa, and Lake cos.	(See vars. below)
S. morrisonii var. *hirtiflorus* Hoffman	+++	Sonoma Co.	E Austin Creek, above Dorr's Cabin
S. morrisonii var. *elatus* Hoffman	+++	Napa/Lake cos. line	$\frac{1}{4}$ mi. W of White's Point, Table Mtn. road, ca. 5 mi. E of Mountain Mill House
S. morrisonii var. *morrisonii*	+++	Sonoma Co.	Big Austin Creek at Layton chromite mine
S. niger Greene	+++	Marin Co.	Tiburon
S. polygaloides Gray	+++	Butte to Fresno cos.	Mariposa Co.: 4.5 mi. S of Coulterville, on road to Bagby
S. tortuosus Kell var. *optatus* Jeps.	+++	Mariposa and Tuolumne cos.	Mariposa Co.: 5.5 mi. S of Coulterville, on road to Bagby
Stylocline amphibola (Gray) J.T. Howell	+++?	Lake to Alameda cos.	Lake Co.: Bartlett Mtn. road, 4 mi. above Lucerne
Tauschia glauca (Coult.& Rose) Math. & Const.	+++	Trinity Co. N to SW Oregon	Del Norte Co.: Stony Creek bog near Gasquet
T. howellii (Coult.& Rose) Macbr.	+++	Del Norte and Siskiyou cos. to SW Oregon	(No Calif. specimens seen)

T. kelloggii (Gray) Macbr.	+++	Coast Ranges, Santa Cruz Co. N to Oregon; Sierra Nevada, Tulare to Butte cos.	Marin Co.: below summit of 602 ft. hill W of Paradise Canyon
Thlaspi montanum L. var. *californicum* (Wats.) P. Holmgren	+++	Humboldt Co. to Josephine Co., Oregon	Humboldt Co.: Kneeland Prairie
Trichostemma rubrisepalum Elmer	+++?	Tuolumne and Mariposa cos.; San Benito Co.	Tuolumne Co.: 2 mi. S of Chinese Camp
Vancouveria chrysantha Greene	+++	Del Norte and Siskiyou cos. to SW Oregon	Del Norte Co.: 0.5 mi. W of Pine Flat Mtn. on Wimer Road
Veronica copelandii Eastw.	+++	Trinity and Siskiyou cos.	Trinity Co.: along trail to Deer Pass from Deer Lake, Trinity Alps
Viola cuneata Wats.	+++?	Mendocino Co. to S Oregon	Humboldt Co.: Horse Mtn.
V. lobata Benth. ssp. *psychodes* (Greene) Munz	+++	Butte Co. to SW Oregon	Del Norte Co.: Stony Creek bog near Gasquet
Zygadenus fontanus Eastw.	++	Mendocino to Marin and San Benito cos.	Del Norte Co.: top of Coon Mtn.

[1] JUB = Jurassic ultrabasic (serpentinite, etc.) rocks; fidelity: +++ = highest (95 - 100%), ++ = high (ca. 85 - 94%), ? = probably endemic to serpentine, but not verified.

[2] Sample localities from specimen label data (WTU, UC, CAS) or personal observation.

Appendix D

Appendix D

Taxa with local or regional indicator status on ultramafic substrates in California

Taxon	1[1]	Overall range	Distribution on ultramafics[2]	Sample localities
FERNS				
Adiantum pedatum L. var. *aleuticum* Rupr.	!	North America, mostly western	North Coast Ranges and Siskiyou Mts. to Wash.	Del Norte Co.: grade from Hurdy Gurdy Creek to Van
Aspidotis densa (Hook.) Lellinger	!!	Cismontane Calif. to British Columbia, E to Quebec	Throughout range	Trinity Co.: Black Lassic; Marin Co.: Rock Springs to Laurel Dell, Mt. Tamalpais
Polystichum imbricans (D.C. Eaton) D.H. Wagner	?	S Calif. to British Columbia	North Coast Ranges and S Oregon	Humboldt Co.: Horse Mtn.
P. kruckebergii W.H. Wagner	?	Western North America	Expected in North Coast Ranges	(None recorded)
P. scopulinum (D.C. Eaton) Maxon	!!	S Calif. to Washington, E to Quebec	North Coast Ranges and SW Oregon	Siskiyou Co.: between Cook-and-Green Pass and Red Butte
GYMNOSPERMS				
Calocedrus decurrens (Torr.)	!!	Lower Calif. to Oregon	Coast Ranges and Siskiyou Mts.	San Benito Co.: Clear Creek summit, 10.1 mi. E of San Benito River
Chamaecyparis lawsoniana (A. Murr.) Parl.	!	NW Calif. to SW Oregon	Nearly throughout range	Del Norte Co.: low divide

Cupressus bakeri Jeps.	!	Shasta and Siskiyou cos.	Burney Springs area of Shasta Co.	Siskiyou Co.: W fork of Seiad Creek
C. bakeri var. ssp. *matthewsii* C.B. Wolf	?	Siskiyou Co.	Same	(No serp. sites recorded)
C. macnabiana A. Murr.	!!	Amador to Butte cos.; Shasta to Sonoma cos.	Throughout range (non-serp. sites rare)	Lake Co.: hills NE of Glenbrook, W end of Cobb Valley
Juniperus communis L. var. *saxatilis* (this may be var. *jackii*)	!	Circumboreal western North America	North Coast Ranges	Del Norte Co.: Gasquet Toll Road, 5-10 mi. NE of Gasquet
Picea breweriana Wats.	!!	Del Norte, Trinity, and Siskiyou cos.	Throughout range	Del Norte Co.: Sanger Lake
Pinus attenuata Lemmon	!!	Cismontane Calif. to S Oregon	Coast Ranges and Siskiyou Mts.	Lake Co.: Socrates Mine, Mayacamas Mts.
P. balfouriana Grev.& Balf. (Klamath Mts. race)	!	Trinity and Siskiyou cos.	Throughout range	Trinity Co.: Deer Pass, Salmon/Trinity Primitive Area
P. coulteri D. Don	?	Contra Costa Co. to lower Calif.	San Luis Obispo, Monterey, and San Benito cos.	San Benito Co.: Clear Creek summit, 10.1 mi. E of San Benito River
P. jeffreyi Grev.& Balf.	!!	Lower Calif. to Oregon	San Benito Co.; Lake to Del Norte cos.; SW Oregon	Siskiyou Co.: Metcalf's Ranch at foot of Mt. Eddy

MONOCOTS

Agropyron trachycaulum (Link.) Malte.	!	North America	South Coast Ranges to Oregon border	Del Norte Co.: Darlingtonia Flats, Smith River

Taxon		Distribution	Locality
Allium cratericola Eastw.	!!	North Coast Ranges and central Sierra Nevada	Trinity to Napa, Butte to Tuolumne cos.	Lake Co.: 2.9 mi. W of Lakeport on road to Hopland
A. lacunosum Wats.	!!	Marin to Santa Barbara cos.	Coast Ranges	San Mateo Co.: Jasper Ridge
A. serratum Wats.	!!	Merced to Colusa cos.	Throughout range	Santa Clara Co.: Colorado Creek, Red Mts.
Brodiaea crocea (Wood) Wats.	!	Trinity Co. to SW Oregon	Throughout range	Siskiyou Co.: E side of Scott Valley
Calochortus coeruleus (Kell.) Wats.	?	Lassen and Tehama to Amador cos.	Throughout range	Tehama Co.: 10 mi. NE of Lomo (serp?)
C. coeruleus (Kell.) Wats. var. *maweanus* Jeps. (= *C. tolmiei* H.& A.)	!	W slope of Sierra Nevada; North Coast Ranges	Del Norte Co.	French Hill, S of Gasquet
C. elegans Pursh. var. *nanus* Wood	?	Siskiyou Co. and Oregon	Throughout range	Siskiyou Co.: Scott Mtn. Public Camp (serp?)
C. invenustus Greene	!!	Tuolumne to Tulare cos.; Santa Clara and Monterey cos.	San Benito Co.	Santa Clara Co.: Colorado Creek, Red Mts., Hamilton Range
C. nudus Wats.	?	Siskiyou to N Eldorado cos.	Coast Ranges	Trinity Co.: Deadfall Meadows, 3 mi. W of Mt. Eddy
C. umbellatus Wood	?	Lake to Santa Clara cos.	Throughout range	Marin Co.: 4 mi. SE of Fairfax
C. uniflorus H.& A.	?	Monterey to Trinity and Mendocino cos.; SW Oregon	Throughout range	Colusa Co.: Bear Valley near Bartlett Springs road

129

Species		Range		Locality
C. vestae Purdy	!	North Coast Ranges	Humboldt to Napa and Sonoma cos.	Lake Co.: near base of SE side of Hill 1030, E of Middletown
Carex mendocinensis Olney	!	Marin to Del Norte cos., and SW Oregon	Throughout range, in seeps	Del Norte Co.: Stony Creek bog near Gasquet
C. serratodens W. Boot	!	Coast Ranges and Sierra Nevada	Expected throughout range	Stanislaus Co.: Del Puerto Canyon
Chlorogalum angustifolium Kell.	!	Shasta to Lake cos.; Fresno Co.	Throughout range	Mendocino Co.: Foothills at S end of Round Valley
Cypripedium californicum Gray	!!	Marin Co. to SW Oregon	Mostly Del Norte Co. and SW Oregon	Humboldt Co.: Horse Mtn.
Festuca tracyi Hitchc.	?	Kern, Colusa, San Benito, Kings, Napa, and Lake cos.	S Coast Ranges	San Benito Co.: 2.6 mi. N of Idria
Fritillaria agrestis Greene	?	Mendocino and San Luis Obispo cos.	Throughout range	(No serp. localities seen)
F. biflora Lindl.	?	Coast Ranges to S Calif.	Coast Ranges	Fresno Co.: Castro Canyon, serp., Jocalitos Creek
F. micrantha Heller	?	Plumas to Tulare cos.	Throughout range	(None seen)
F. phaeanthera Eastw.	?	Butte and Napa cos.	Throughout range	(None seen)
F. pluriflora Torr. in Benth.	?	Mendocino, Glenn, Solano, and Butte cos. to S Ore.	Throughout range	(No serp. localities seen)
F. recurva Benth. var. recurva	!	Lake Co. N to Oregon; Nevada Co. W to Nevada	Coast Ranges	Trinity Co.: on Red Lassic

Species		Distribution	Range	Locality
Melica stricta Bol.	?	Western North America	North Coast Ranges and New Idria area	San Benito Co.: San Benito Mtn.
Muilla maritima (Torr.) Wats.	!	Glenn to San Diego cos.; Sierra foothills	Coast Range serpentines	San Mateo Co.: Spring Valley, Crystal Springs Lake
Odontostomum hartwegii Torr.	!	Sierra foothills and Coast Ranges	Butte to Mariposa and Tehama, and Napa cos.	Eldorado Co.: serpentine plateau between Green Valley and Eggars
Panicum thermale Bol.	!	Sonoma Co. N to Washington, etc.	Thermal areas of North Coast Ranges	Lake Co.: road from Middletown to Peterson's
Poa tenerrima Scribn.	!	Eldorado and Ventura cos.	Sierra foothills; serpentine seeps	Eldorado Co.: near road, 1.7 mi. SW of Garden Valley
Xerophyllum tenax (Pursh) Nutt.	!	California to Washington	N Sierra Nevada, North Coast Ranges	Humboldt Co.: Horse Mtn.
DICOTS				
Acanthomintha ilicifolia (Gray) Gray	?	S California (and San Francisco Bay Region?)	San Mateo Co.?	San Mateo Co.: serp. near Menlo Park (misidentified?)
A. obovata Jeps. ssp. *obovata*	!	San Benito to Ventura cos.	Throughout range	San Benito Co.: New Idria Mine
Aconitum columbianum Nutt. Named variants:				
A. hanseni Greene	?	Eldorado, Mariposa cos.	Throughout range	(None seen)
A. viviparum Greene	?	Del Norte, Siskiyou cos.	Throughout range	(None seen)
A. leibergii Greene	?	Humboldt, Trinity cos.	Throughout range	(None seen)
A. geranioides Greene	?	Humboldt to Glenn, Siskiyou, and Plumas cos.; S Oregon	Throughout range	(None seen)

131

Angelica tomentosa Wats.	!!	San Francisco Bay Region to SW Calif.	Throughout range	Lake Co.: St. Helena Creek
Antirrhinum breweri Gray	!!	Sonoma and Mariposa cos. to SW Oregon	Throughout range	Trinity Co.: trail to Globe Stamp Mill, Dedrick, Canyon Creek
A. cornutum Benth.	!!	North Coast Ranges and N Sierra Nevada	Eldorado, Napa, Humboldt, and Shasta cos.	Lake Co.: meadow on Pope Valley/Middletown road, 3 mi. NW of county line
A. vexillo-calyculatum Kell.	!!	Sonoma to San Benito cos.	Expected throughout range	Trinity Co.: 10 mi. SE of Peanut, between Hayfork and Wildwood
Aquilegia eximia Van Houtte ex Planch.	!!	North and south Coast Ranges	Seeps, San Benito to Mendocino cos.	Trinity Co.: SW of Peanut (may be serp. endemic)
Arabis subpinnatifida Wats.	!!	North Coast Ranges to SW Oregon	Probable sites in Humboldt, Glenn, and Siskiyou cos.	Siskiyou Co.: 4 mi. E of Etna, E side of Callahan road, edge of Weston Gulch
A. suffrutescens Wats.	!	North Coast Ranges, Sierra Nevada to Wash. and Idaho	Yolla Bolly Mts. and Siskiyou Co.	Siskiyou Co.: peak above Siskiyou Mt. divide, Takilma/Happy Camp road

Arctostaphylos (several spp. may be local or regional indicators—e.g., *auriculata, candidissima, canescens, columbiana, densiflora, elegans, glandulosa, glauca, manzanita, montana* (or endemic?), *nevadensis, patula, sensitiva, stanfordiana, virgata*)

Species		Distribution	Serpentine range	Locality
A. stanfordiana Parry ssp. stanfordiana	?	Napa, Lake, Sonoma, and Mendocino cos.	Throughout range	Lake Co.: along road across canyon from Oat Hill Mine
A. viscida Parry	!	Sierra Nevada: Kern Co.; North Coast Ranges: Napa and Lake cos.	Throughout range?	(No certain serp. localities seen)
Arenaria douglasii Fenzl. ex. T.& G.	!!	Oregon to lower Calif.	North Coast Ranges and Sierra Nevada	Siskiyou Co.: along road no. 8, 2.4 mi. W of jctn. with State Rte. 3
A. nuttallii Pax ssp. gregaria (Heller) Maguire	!	Lake to Siskiyou cos.	Throughout range	Trinity Co.: Barren mtn. top R6E T1S, sec. 31
Astragalus curtipes Gray	!	Coastal and south Coast Ranges	San Luis Obispo and San Benito cos.	San Luis Obispo Co.: 1.0 mi. N of San Luis Obispo
A. rattanii Gray	!	Humboldt to central Mendocino and NW Lake cos.	Throughout range	Humboldt Co.: Kneeland Prairie
Berberis piperiana (Abrams) McMinn	!	Coast Ranges from Lower Calif. to S Oregon	Lake to Del Norte and Siskiyou cos.	Siskiyou Co.: N base of Mt. Eddy
B. pumila Greene	!!	North Coast Ranges and Sierra Nevada	Lake to Siskiyou, and Mariposa cos.	Siskiyou Co.: $\frac{1}{2}$ mi. S of W branch Indian Creek road, Happy Camp/Takilma road
Calycadenia ciliosa Greene	?	Butte and Lake cos. N to S Oregon	Throughout range	N Lake Co.: Hot dry slopes "Pine Mt.," Elk Mtn. (serp.)
C. hispida (Greene) Greene ssp. hispida	?	Placer Co. to Kern Co. and S Monterey Co.	Monterey Co.?	(None seen)

C. hispida ssp. *reducta* Keck	?	Contra Costa to Santa Clara cos.	Throughout range	(None seen)
C. mollis Gray	?	Tuolumne to Tulare cos.	Throughout range	(None seen)
C. multiglandulosa DC ssp. *bicolor* (Greene) Keck	?	Butte to Tulare cos.	Throughout range	(None seen)
C. multiglandulosa DC ssp. *cephalotes* (DC) Keck	!	Coast Ranges	Mendocino to San Mateo cos.	San Mateo Co.: Jasper Ridge
C. multiglandulosa ssp. *robusta* Keck	!!	Santa Clara and adjacent San Francisco Bay area cos.	Throughout range	Marin Co.: Mt. Tamalpais
C. oppositifolia (Greene) Greene	!!	Butte Co.	Butte Co.	3 mi. E of Pentz, Feather River
C. spicata (Greene) Greene	?	Butte to Tulare cos.	Throughout range	(None seen)
C. truncata ssp. *microcephala* Hall ex Keck	?	Trinity to Lake cos. local in Santa Lucia Mts. of S Monterey Co.	Throughout range	(None seen)
C. truncata ssp. *scabrella* (E. Drew) Keck	!	Interior foothills from Placer and Lake cos. to S Oregon	Throughout range	Plumas Co.: 8.6 mi. SW of Arch Rock tunnel, N Fork of Feather River
C. truncata DC ssp. *truncata*	!	Sierra Nevada foothills; inner Coast Ranges	Plumas to Mariposa and Del Norte to Santa Clara cos.	Del Norte Co.: Waldo/ O'Brien area
Calycanthus occidentalis H.&A.	?	North Coast Ranges and Sierra Nevada	Napa to Trinity and Lake Tulare to Shasta cos.	Lake Co.: Putah Creek drainage, Middletown

Species	Range	Distribution	Locality
Calyptridium quadripetalum Wats. !!	Lake to Glenn cos.	Throughout range	Tehama Co.: SW ridge leading to Eagle Peak
C. umbellatum (Torr.) Greene ! (= *Spraguea umbellata* Torr.)	Western North America	North Coast Ranges	Siskiyou Co.: on road to Bolan Lake, 3 mi. from Takilma/Happy Camp road
Calystegia fulcrata (Gray) Brummitt (fide G.L. Stebbins) ?	Eldorado Co.	Eldorado Co.	(None seen)
C. stebbinsii Brummitt ?	Nevada and Eldorado cos., mostly on gabbro	Within range	Eldorado Co.: U.S. Hwy 50, 18 mi. W of Placerville (gabbro or serp.?)
Camissonia lacustris Raven !!	Coast Ranges and Sierra Nevada	Coast Ranges (fide P. Raven)	(None seen)
Campanula scabrella Engelm. ?	Klamath/Siskiyou area N to Wash. and Idaho	Siskiyou Co.	Siskiyou Co.: Cory Peak area
Castilleja foliolosa H.& A. !!	Coast Ranges S to lower Calif. and Sierra Nevada	Bay Area N in Coast Ranges	Napa Co.: Hunting Creek, Napa/Lake cos. line
C. pruinosa Fern. !?	North Coast Ranges and Sierra Nevada	Del Norte, Siskiyou, and Tuolumne cos.	Humboldt Co.: Horse Mtn.
C. stenantha Gray !!	S Calif. to Lake and Fresno cos.	Central and N Coast Ranges and central Sierra Nevada	Tuolumne Co.: 1½ mi. SW of Chinese Camp
Ceanothus foliosus Parry var. *medius* McMinn !!	Santa Clara to San Luis Obispo cos.	Within range	San Luis Obispo Co.: 4 mi. NW of Cuesta Summit
C. papillosus T.& G. var. *roweanus* McMinn !!	San Benito and Monterey cos.; Santa Ana Mts.	In both segments of range	Orange Co.: Pleasant Peak, Santa Ana Mts.

Chaenactis glabriuscula DC var. *gracilenta* (Greene) Keck	!!	Colusa and Mendocino to San Benito cos.	Throughout range	Napa Co.: Knoxville
C. glabriuscula DC (other vars. than *gracilenta*	!	Coast Ranges and Sierra Nevada	Throughout range	Lake Co.: serp. hills 2 mi. S of Lakeport
Chorizanthe palmeri Wats.	!!	San Benito to Santa Barbara cos.	Throughout range	Santa Barbara Co.: Figueroa Mtn. road just S of Grass Mtn.
Clarkia arcuata (Kell.) Nels. & Macbr.	!!	Butte to Mariposa cos.	Throughout range	Mariposa Co.: between Coulterville and Bagby
Collinsia sparsiflora F.&M.	!!	Coast Ranges and Sierra Nevada	Contra Costa to Humboldt, and Tuolumne to Butte cos.	Marin Co.: Matt Davis trail, Mt. Tamalpais
Convolvulus malacophyllus Greene	!!	North Coast Ranges and Sierra Nevada	Stanislaus to Trinity and Siskiyou cos.; Tulare Co.	Siskiyou Co.: near Wagon Creek Falls E side of Mt. Eddy
Cordylanthus pilosus Gray subsp. *pilosus* (serp. race)	!!	Lake to Siskiyou cos.	Throughout range	Lake Co.: along Hwy 29, just beyond Napa Co. line
C. tenuis Gray ssp. *tenuis*	!!	Sierra Nevada	Throughout range	Eldorado Co.: 2.4 mi. S of Georgetown on Marshall Rd. (serpentine?)
C. tenuis ssp. *viscidus* (Penn.) Heckard	!!	North Coast Ranges to SW Oregon	Tehama and Trinity to Del Norte cos.	Humboldt Co.: 8 mi. S of Hoopa Valley on new road to Trinity Summit

Species		Distribution	Range	Locality
Coreopsis stillmanii (Gray) Blake	!!	Butte to Tulare, and Contra Costa to Stanislaus cos.	Throughout ranges in Sierras and Coast Ranges	Tuolumne Co.: 4 mi. N of Coulterville
Corethrogyne californica DC	!!	San Francisco to Monterey Peninsula	Throughout range	San Mateo Co.: E side of Crystal Springs Lake
Cryptantha excavata Bdg.	?	Yolo to Colusa and Lake cos.	Throughout range	(None seen)
C. milobakeri Jtn.	!!	Lake to Del Norte, W Siskiyou and Plumas cos.; S Oregon	Throughout range	Colusa Co.: John Smith Road (Goat Mtn./Stonyford road)
C. nemaclada Greene	?	Colusa to San Luis Obispo cos.; Tehachapi Mts.	Coast Ranges	San Benito Co.: 10 mi. E of Paicines (serp?)
C. sparsiflora (Greene) Greene	?	W Stanislaus and San Benito cos.; Mariposa to Kern cos.	Most of range	Fresno Co.: 2 mi. above Dunlap on Sand Creek cutoff to Miramonte
C. subretusa Jtn.	!!	Siskiyou Co., Oregon, and N Nevada	Siskiyou Mts. area	Trinity Co.: W and S slopes of Mt. Eddy
Darlingtonia californica Torr.	!	North Coast Ranges to SW Oregon; N Sierra Nevada	Del Norte, Trinity, Siskiyou, and Plumas cos.	Del Norte Co.: S fork of Smith River trail near Rock Creek Lodge
Delphinium californicum T.&G. ssp. *interius* (Eastw.) Ewan	?	Contra Costa to Santa Clara cos.	Throughout range	San Mateo Co.: Crystal Springs Lake (non var. *interius*?)
D. decorum F.&M. ssp. *tracyi* Ewan	?	Lake to Humboldt cos.; S Oregon	Throughout range	Tehama Co.: S Yolla Bolly Mtn. (serp?)

Species		Distribution	General range	Specimens seen
D. gypsophilum Ewan ssp. *parviflorum* Lewis & Epling	?	Monterey and San Luis Obispo cos.	Throughout range	(None seen)
D. hesperium Gray ssp. *hesperium*	!	Humboldt and Butte cos. to Santa Clara Co.	Throughout range	San Mateo Co.: serp. back of Menlo Park
D. hesperium ssp. *pallescens* (Ewan) Lewis & Epling	?	Tehama to Santa Clara and Kern cos.	Coast Ranges	Napa Co.: along Eticura Creek, S of Knoxville (serp?)
D. patens Benth. ssp. *patens*	?	Lake and Colusa cos. to N Santa Barbara Co.; Butte to Eldorado Co.	Throughout range	(None seen)
D. recurvatum Greene	?	Contra Costa to Kern cos.; Glenn and Butte cos.	North Coast Ranges	(None seen)
D. sonnei Greene	!	Del Norte and Siskiyou cos.; Eldorado Co.; Oregon; Nevada	North Coast Ranges	Trinity Co.: Deadfall Mds., 3 mi. W of Mt. Eddy
D. umbraculorum Lewis & Epling	?	Monterey and San Luis Obispo cos.	Throughout range	(None seen)
D. variegatum T.&G.	?	San Luis Obispo and Tulare cos. to Tehama Co.	Most of range	Napa Co.: Samuel Spr., 7 mi. NW of Monticello
Dentaria californica Nutt. var. *cuneata* (Greene) Detl.	?	San Benito and Monterey cos.	Throughout range	San Benito Co.: along Tres Pinos Cr., 1 mi. SE of Emmet School, Panoche Pass road
D. tenella Pursh. var. *palmata* Detl.	?	Siskiyou and Trinity to Placer cos.	Throughout range	(None seen)

Species		Distribution	Range	Locality
Dicentra pauciflora Wats.	!	Salmon/Trinity Alps and Sierra Nevada	Salmon/Trinity Alps	Siskiyou Co.: road to Bald Mtn., ca. 6 airline mi. W of Happy Camp
Dodecatheon clevelandii Greene ssp. *patulum* (Greene) H.J. Thomps.	!!	Tehama to Kern cos.; San Francisco Bay Region to San Benito Co.	Seeps throughout range	Alameda Co.: Arroyo Mocho, 8 mi. from S Livermore road on Mines road (serp.?)
Draba howellii Wats.	!!	North Coast Ranges	Salmon/Trinity Alps, Marble and Siskiyou Mts.	Siskiyou Co.: Young's Valley near Cyclone Gap
Dudleya blochmaniae (Eastw.) Moran	!!	San Luis Obispo Co. to lower Calif.	Northern part of range	(None seen)
D. cymosa (Lem.) Ritt & Rose	?	Amador and Calaveras cos.	Throughout range	(None seen)
D. cymosa ssp. *setchellii* (Jeps.) Moran	?	Contra Costa Co. to the Pinnacles, San Benito ccs.	Throughout range	(None seen)
Epilobium minutum Lindl. ex Hook. var. *foliosum* T.&G.	!!	Widespread in Calif., N to British Columbia	Coast Ranges	Sonoma Co.: Gilliam Creek, a tributary of E Austin Creek
E. niveum Bdg.	?	Lake and Mendocino cos.	Throughout range	Lake Co.: Snow Mtn. (serp.?)
E. obcordatum Gray var. *laxum* (Hausskn.) Dempst. ex Jeps.	?	Siskiyou, Trinity, and Placer cos.	Likely throughout range	(None seen)
E. oreganum Greene	?	Klamath Ranges and Sierra Nevada	Klamath Ranges	(None seen)

Species	Range		Locality
Eriogonum ovilleanum ssp. *adsurgens* (S. Stokes) Abrams !!	Stanislaus to San Luis Obispo cos.	Throughout range	Stanislaus Co.: Arroyo del Puerto, Red Mts., Mt. Hamilton Range
E. pyrolaefolium Hook. !	N Calif. to Wash. and Montana	Siskiyou, Trinity, and Del Norte cos.	Siskiyou Co.: (Mt. Eddy?)
E. umbellatum Torr. ssp. *bahiaeforme* (T.&G.) Munz !	North Coast Ranges to S Calif.	Coast Ranges from Siskiyou to San Benito cos.	San Benito Co.: Mt. San Carlos, near Idria
Eriophyllum confertiflorum (DC) Gray ?	Cismontane central and S Calif.	South Coast Ranges?	San Benito Co.: 1 mi. S of Idria
E. confertiflorum var. *tanacetiflorum* (Greene) Jeps. !!	Calaveras and Mariposa cos.	Throughout range	Mariposa Co.: 2.2 mi. N of Bagby on Hwy 49
E. lanatum (Pursh) Forbes var. *achillaeoides* (DC) Jeps. !	Klamath Mts. to Santa Clara and Santa Cruz cos.; Mariposa Co.	Coast Ranges	Lake Co.: between Middletown and Pope Valley
E. lanatum var. *aphanactis* J.T. Howell ?	Glenn, Colusa, and Lake cos.	Throughout range	(None seen)
E. lanatum var. *grandiflorum* (Gray) Jeps. ?	Mariposa to Shasta cos.; Mendocino to Siskiyou and and Del Norte cos.; SW Oregon	North Coast Ranges; SW Oregon	Trinity Co.: beside Hwy 299 just E of Big Bar (serp.?)
E. lanatum var. *lanceolatum* (Howell) Jeps. !	Klamath Mts.; SW Oregon	Throughout range	Humboldt Co.: Klamath River 12 mi. N of Weitchpec
E. nubigenum Greene var. *congdonii* (Bdg.) Const. ?	Mariposa Co.	Within range	(None seen)
Erysimum franciscanum G. Rossb.	San Mateo to Marin cos.	Throughout range	San Mateo Co.: below Crystal Springs Lake

Species		Range	Distribution	Locality
Frasera umpquaensis Peck & Applegate	!	Trinity Co. to SW Oregon	Throughout range	Trinity Co.: Horse Ridge?
Fremontodendron californicum Cov. var. *napensis* McMinn	!	Napa, Lake, and Yolo cos.	Throughout range	Lake Co.: 3 mi. W of Knoxville
F. californicum ssp. *obispoensis* (Eastw.) Munz	!!	San Luis Obispo and N Santa Barbara cos.	Throughout range	San Luis Obispo Co.: Cuesta road NW of U.S. Hwy 101
F. decumbens R. Lloyd	?	El Dorado Co.	Within range	El Dorado Co.: olivine-gabbro ridge S of Pine Hill
Galium ambiguum Wight	?	Lake and Tehama cos. to S Oregon	Throughout range	Shasta Co.: 1 mi W of Regan Meadow Camp
Gentiana newberryi Gray	?	Sierra Nevada, Siskiyou Mts., W Nevada, and S Oregon	Local with *Darlingtonia* in Amador Co.	(None seen)
Githopsis specularioides Nutt.	?	Cismontane Calif. N to Washington	North Coast Ranges?	
Gutierrezia californica (DC) T.& G.	!	Widespread in Cismontane Calif. E to Arizona	Coast Ranges?	San Francisco Bay Region: Angel Island, etc.
Haplopappus squarrosus H.& A. ssp. *stenolepis* Hall	!!	Fresno to Santa Barbara cos.	Throughout range	Monterey Co.: S side of Parkfield Grade
Hemizonia halliana Keck	!!	San Benito to San Luis Obispo cos.	Throughout range	San Benito Co.: 6 mi. from New Idria on road to Panoche

Hesperolinon breweri (Gray) Small	!!	Central Coast Ranges	Napa Co.: S end of Cappell Valley
H. californicum (Benth.) Small	!!	Glenn and Butte cos. to San Carlos Range	Napa Co.: W side Chiles Valley, road from Rutherford to Monticello
H. clevelandii (Greene) Small	!!	Santa Clara to Mendocino cos.	Stanislaus Co.: Red Mtns., Mt. Hamilton Range
H. micranthum (Gray) H.K. Sharsm.	!	Coast Ranges and Sierras to S Oregon	Lake Co.: just E of Black Oak Villa, Butts Canyon
Ivesia gordonii (Hook) T.& G.	!	Trinity, Humboldt, and Tuolumne cos. to Rocky Mts.	Siskiyou Co.: Mt. Eddy, at 9000 ft.
I. pickeringii Torr. ex Gray	?	Siskiyou and Trinity cos.	(None seen)
Layia septentrionalis Keck	!	Colusa to Sonoma cos.	Lake Co.: 2.5 mi. S of Lakeport
Lewisia cotyledon (Wats.) Robins. in Gray ssp. *cotyledon*	!	NW Calif. to SW Oregon	Siskiyou Co.: Twin Valley train near El Capitan
L. cotyledon ssp. *heckneri* (Mort.) J. Hohn <u>in ed.</u>	!!	Trinity Co.	Trinity Co.: along trail, Canyon Creek
L. cotyledon ssp. *howellii* (Wats.) J. Hohn <u>in ed.</u>	!!	NW Calif. to S Oregon	Humboldt Co.: mouth of Bluff Creek, 6 mi. NE of Weitchpec

Species		Range	Region	Locality
L. leana (Porter) Rob. in Gray	!!	Sierra Nevada and NW Cal f. to SW Oregon	Klamath-Siskiyou Mts.	Siskiyou Co.: on road to Bolan Lake, 3 mi. from Takilma/Happy Camp road
L. oppositifolia (Wats.) Rob in Gray	!!	Del Norte Co. to SW Oregon	Throughout range	Del Norte Co.: Gordon Mtn., on road N of Big Flat
L. rediviva Pursh.	!	Western North America	Central Coast Ranges	Lake Co.: above meadow 2.9 mi. W of Lakeport on road to Hopland
L. triphylla (Wats.) Rob.	!	Western North America	Sierra Nevada (?) and north Coast Ranges	Colusa Co.: Cooks Springs
Linanthus liniflorus (Benth.) Greene	!	Contra Costa to San Luis Obispo cos.	Throughout range	San Francisco Co.: tip of Hunter's Point
Linum perenne L. ssp. *lewisii* (Pursh) Hult.	!	Western North America	South Coast Ranges	San Benito Co.: San Benito Mtn.
Lomatium parvifolium (H.&A.) Jeps.		Monterey to San Luis Obispo cos.	Throughout range	San Luis Obispo Co.: Serrano Canyon near San Luis Obispo
L. repostum (Jeps.) Math.	!!	Napa and Lake cos.	Throughout range	Lake Co.: 7-8 mi. from Lower Lake, on road to Knoxville
Mimulus brachiatus Penn.	?	Lake Co.	Throughout range	(Not seen)
M. douglasii (Benth. in DC) Gray	!!	San Benito and Tulare cos. N to Oregon	Throughout range	Santa Clara Co.: divide between Colorado Creek and San Antonio Creek

M. nasutus Greene (= *M. pardalis* of Abrams, form of *M. guttatus*)	!!	Western North America	Sierra Nevada	Tuolumne Co.: Peoria Flat
Montia gypsophyloides (F.& M.) Howell	!!	San Luis Obispo to Mendocino cos.	Throughout range	Sonoma Co.: Harrison Grade between Occidental and Camp Meeker
Navarretia pubescens (Benth.) H.& A.	!!	Humboldt and Butte cos. to San Luis Obispo and Kern cos.	Throughout range	Lake Co.: Meadow, Locomono Creek, E of Middletown
Orthocarpus copelandii Eastw.	!!	Glenn to Siskiyou cos., SW Oregon	Throughout range	Trinity Co.: Scott Mtn.
O. lacerus Benth.		Fresno to Modoc and Siskiyou cos., S Oregon	Sierra Nevada, W slope	Amador Co.: just E of site of Waits Station, along Rte. 16, 2 mi. W of jct. with Rte. 49
Parvisedum pumilum (Benth.) Clausen		Napa Range and Sierra Nevada foothills	Throughout range	Mariposa Co.: 9 mi. SE of Coulterville
Perideridia oregana (Wats.) Math.	!	Glenn and Trinity cos. N to Wash.	Coast Ranges	Del Norte Co.: Old Gasquet toll road. 0.5 mi. above bridge over Smith River just N of Gasquet
P. pringlei (Coult. & Rose) Nels. & McBr.	!!	Los Angeles to San Luis Obispo and Kern cos.	San Luis Obispo Co.	San Luis Obispo Co.: CalPoly College Botanical Garden
Phacelia californica Cham.	!!	Santa Clara to Del Norte cos.	Throughout range	Marin Co.: Tiburon

Species		Distribution	Range	Locality
P. divaricata (Benth.) Gray	!	Mendocino to San Benito and Monterey cos.; Mariposa to Kern cos.	Coast Ranges	Marin Co.: Alpine Lake road, 2 mi. SW of Fairfax
P. egena (Greene ex Brand.) Const.	!!	Kern to Siskiyou cos.; Tehama and Glenn cos.	North Coast Ranges	Tehama Co.: Covelo road 5 mi. W of Paskenta
P. imbricata Greene subsp. *imbricata*	!	S Calif. to Shasta and Humboldt cos.	Mostly north Coast Ranges	Lake Co.: Hill 1030, 3 mi. N of Middletown
P. leonis J.T. Howell	?	Trinity and Siskiyou cos.	Throughout range	Siskiyou Co.: Twin Valley 4 mi. NNE of Preston Peak
P. pringlei Gray	!!	Trinity and Siskiyou cos.	Throughout range	Trinity Co.: serp. N slope between Coffee and Eagle creeks, 6 mi. N of Carrville
Plagiobothrys greenei (Gray) Jtn.	?	Coast Ranges and Sierras	Coast Ranges	Lake Co.: 5 mi. W of Lower Lake (serp.?)
P. infectivus Jtn.	?	Colusa to San Luis Obispo cos.	Throughout range	Fresno Co.: 13.4 mi. W of Coalinga (serp.?)
P. lithocaryus (Greene ex Gray) Jtn.	?	Lake and Mendocino cos.	Throughout range	(None seen)
P. shastensis Greene ex Gray	?	Stanislaus to Butte cos.; Merced to Siskiyou cos.	Throughout range	(None seen)
Polemonium chartaceum Mason	!	Siskiyou and Mono cos.	Klamath/Siskiyou Mts.	Siskiyou Co.: Dobbins Lake, Mt. Eddy

Species		Range	Region	Locality
Psoralea californica Wats.	!	Glenn Co. to S Calif.	Inner Coast Ranges	San Benito Co.: at turnoff to Picacho Mine, ca. 6 mi. S of Hernandez
Pyrola picta Sm. subsp. *dentata* (Sm.) Piper	!	Mendocino Co. to British Columbia	Klamath/Siskiyou Mts.	Siskiyou Co.: 15 mi. N of Happy Camp on Takilma road
Quercus vaccinifolia Kell.	!!	Sierra Nevada and SW Calif.	Klamath/Siskiyou Mts.	Trinity Co.: Red Lassic area
Raillardella scabrida Eastw.	?	Lake to Mendocino and Tehama cos.	Throughout range	Trinity Co.: Black Lassic (serp.?)
Rhamnus californica Esch. var. *tomentella* (Benth.) C.B. Wolf.	!	Cismontane Calif.	Inner Coast Ranges	Sonoma Co.: 2 mi. NNE of Occidental
Salix coulteri Anderss.	!!	San Luis Obispo Co. N to Washington	San Luis Obispo Co. and nearby areas	San Luis Obispo Co.: Serrano Canyon
Salvia sonomensis Greene	!	Sierra Nevada and Coast Ranges	South Coast Ranges N to Napa and Lake cos.	Lake Co.: Kelseyville, lower Lake area
Sedum obtusatum Gray subsp. *obtusatum*	!!	Sierra Nevada from Tulare to Plumas cos.; north Coast Ranges	Throughout range	Trinity Co.: W slopes of Mt. Eddy, just above the Deadfall Lakes
Senecio eurycephalus T.& G.	!	Sonoma and Colusa to Siskiyou, Shasta, Butte, Lassen, and Modoc cos.; Oregon	North Coast Ranges	Napa Co.: along Morgan Valley road, 1.5 mi. E of Lake Co. line
Silene californica Dur.	?	Widespread in cismontane Calif.	Local indicator in north Coast Ranges?	Del Norte Co.: Low Divide

Species		Range		Locality
S. campanulata Wats. subsp. *greenei* (Wats.) Hitchc. & Maguire	!!	Del Norte, Humboldt, Trinity, and Shasta cos.	Throughout range	Trinity Co.: road to Peanut, 3 mi. beyond Wildwood
S. grayi Wats.	!!	Siskiyou and Trinity cos.	Throughout range	Siskiyou Co.: peak above Siskiyou divide, Takilma/Happy Camp road
S. hookeri Nutt. ex T.&G. var. *hookeri*	!!	Trinity, N Humboldt, Siskiyou, and Del Norte cos.; S Oregon	Throughout range in Calif.	Del Norte Co.: French Hill 2 mi. S of Gasquet
Streptanthus albidus Greene	!!	Santa Clara, Contra Costa, and Alameda cos.	Throughout range	Santa Clara Co.: 1 mi. E of Hwy 101 on Metcalf Road
S. glandulosus Hook. ssp. *glandulosus*	!!	Sonoma and Solano to San Luis Obispo cos.	Throughout range	Napa Co.: 2.2 mi. NE of Aetna Springs on raod to Middletown
S. glandulosus ssp. *secundus* (Greene) Kruckeberg	!!	Mendocino to Marin cos.	Throughout range	Sonoma Co.: NW of Calistoga, 6.2 mi. W of Hwy 128 on Franz Valley road
Tauschia glauca (Coult. & Rose) Math. & Const.	!!	Trinity Co. N to SW Oregon	Throughout range	Del Norte Co.: Stony Creek bog near Gasquet
T. hartwegii (Gray) Macbr.	!!	S Calif. to Contra Costa and Butte cos.	South Coast Ranges	San Benito Co.: 1.5 mi. SE of San Benito/Hernandez Hwy on road to Laguna Ranch, San Carlos Range

Taxon	Indicator status[1]			Location
Thelypodium flavescens (Hook.) Wats.	!!	Solano to San Benito cos.	Throughout range	Contra Costa Co.: 2 mi. inside old N entrance to Mt. Diablo State Park
Thlaspi montanum L. var. *montanum*	!!	Widespread in western U.S.	North Coast Ranges	Humboldt. Co.: Klamath River 7 mi. NE of Weitchpec
Umbellularia californica (H.&A.) Nutt. (undescribed shrub form)	!!	Cismontane Calif. to SW Oregon	Central and north Coast Ranges	San Mateo Co.: Jasper Ridge
Vancouveria planipetala Calloni	!!	Monterey Co. to SW Oregon	North Coast Ranges	Del Norte Co.: confluence of Shelley and Patrick Creeks off Patrick Creek road
Viola purpurea Kell. ssp. *mohavensis* Baker & Clausen	!	S Calif. to central Coast Ranges	Coast Ranges	Stanislaus Co.: Adobe Creek, Red Mts., Mt. Hamilton Range
V. ocellata T.& G.	!!	Monterey to Del Norte cos.	Throughout range	(None seen)

[1] = Indicator status: !! = high local or regional fidelity to serpentinite; ! = medium fidelity; ? = probable to possible indicator.

[2] Serpentinite, peridotite, and other ultramafic outcrops occur within areas designated; "throughout range" = ultramafic rocks are found within overall range of taxon.

Appendix E

Appendix E
Some serpentine endemic and indicator taxa[1]
listed by geographic province

I. South Coast Ranges (Santa Barbara to Santa Clara cos.)

Serpentine endemics	Serpentine Indicators

Acanthomintha lanceolata
Allium fimbriatum var. *diabolense*
A. fimbriatum var. *sharsmithiae*
A. howellii var. *sanbenitensis*
Arctostaphylos obispoensis
Arenaria douglasii var. *emarginata*
Benitoa occidentalis
Calochortus clavatus
C. obispoensis
Camissonia benitensis
Carex obispoensis
Ceanothus ferrisae
Chorizanthe breweri
C. uniaristata
Cirsium campylon
C. fontinale var. *obispoense*
Delphinium parryi var. *eastwoodiae*
Dudleya abramsii ssp. *murina*
D. bettinae
Emmenanthe penduliflora var. *rosea*
Eriogonum argillosum
E. covilleanum
Eriophyllum jepsonii
Fritillaria falcata
Galium andrewsii var. *gatense*
G. hardhamiae
Hemizonia halliana
Hesperolinon disjunctum
Layia discoidea
Lessingia ramulosa var. *glabrata*

Acanthomintha obovata
Allium lacunosum
Arenaria douglasii var. *emarginata*
Astragalus curtipes
Calycadenia hispida
Ceanothus foliosus var. *medius*
C. papillosus var. *roweanus*
Chorizanthe breweri
C. palmeri
C. uniaristata
Cryptantha nemaclada
Delphinium gypsophilum
  ssp. *parviflorum*
Dentaria californica var. *cuneata*
Dudleya blochmaniae
Emmenanthe penduliflora var. *rosea*
Eriogonum argillosum
E. covilleanum
E. covilleanum ssp. *adsurgens*
Fremontodendron californica
  ssp. *obispoensis*
Haplopappus squarrosus
  ssp. *stenolepis*
Hemizonia halliana
Lessingia ramulosa var. *glabrata*
Linanthus liniflorus
Lomatium parvifolium
Perideridia pringlei
Pinus coulteri
Streptanthus albidus

151

Linanthus ambiguus
Monardella benitensis
M. palmeri
Navarretia mitracarpa
  ssp. jaredii
Phacelia breweri
P. phacelioides

Sidalcea hickmanii **var.** anomala
Streptanthus amplexicaulis
  var. barbarae
S. insignis
S. insignis **ssp.** lyonii
Tauschia hartwegii
Viola purpurea **ssp.** mohavensis

II. Bay Area (San Francisco, Marin, San Mateo, Contra Costa, and Alameda cos.)

Serpentine endemics

Acanthomintha obovata **ssp.** duttonii
Arctostaphylos hookeri
  ssp. franciscana
Calochortus tiburonensis
Castilleja neglecta
Cirsium fontinale
C. vaseyi
Clarkia franciscana
Cordylanthus nidularius
Eriogonum caninum
Eriophyllum latilobum
Hesperolinon congestum
Lessingia ramulosa **var.** micradenia
Montia spathulata **var.** rosulata
Navarretia heterodoxa **ssp.** rosulata
Sidalcea hickmanii **ssp.** viridis

Streptanthus batrachopus
S. glandulosus **ssp.** pulchellus
S. niger
Tauschia kelloggii

Serpentine indicators

Acanthomintha ilicifolia
Ceanothus foliosus **var.** medius
Corethrogyne californica
Eriogonum caninum
Erysimum franciscanum
Gutierrezia californica
Lessingia ramulosa
  var. micradenia

III. North Bay Counties (Napa, Sonoma, and Lake cos.)

Serpentine endemics

Allium fimbriatum Wats. var. *purdyi*

Arctostaphylos stanfordiana
  ssp. *bakeri*

Calamagrostis ophitidis

Calycadenia pauciflora

Ceanothus jepsonii var. *albiflorus*

Cordylanthus pringlei

C. tenuis ssp. *brunneus*

C. tenuis ssp. *capillaris*

Cryptantha clevelandii var. *dissita*

C. hispidula

Delphinium uliginosum

Erythronium helenae

Hesperolinon bicarpellatum

H. drymarioides

H. spergulinum

Madia hallii

Mimulus nudatus

Monardella villosa ssp. *neglecta*

M. viridis

Navarettia jepsonii

Nemacladus montanus

Senecio clevelandii

Streptanthus brachiatus

S. hesperidis

S. morrisonii ssp. *hirtiflorus*

S. morrisonii ssp. *elatus*

S. morrisonii ssp. *morrisonii*

Serpentine indicators

Epilobium foliosum (= *E. minutum*
  var. *foliosum*

E. niveum?

Fremontodendron californicum
  ssp. *napensis*

Githopsis specularioides

Hesperolinon breweri

Layia septentrionalis

Lomatium repostum

Mimulus brachiatus

Rhamnus californica
  var. *tomentella*

Streptanthus glandulosus
  ssp. *secundus*

IV. North Coast Ranges (N Lake and Colusa cos. N to Oregon border)

Serpentine endemics   Serpentine indicators

Allium hoffmanii   *Adiantum pedatum* var. *aleuticum*

Antennaria suffrutescens   *Arabis subpinnatifida*

Arabis aculeolata   *Arenaria douglasii* ssp. *gregaria*

A. macdonaldiana   *Astragalus rattanii*

A. serpentinicola   *Berberis piperiana*

Arctostaphylos stanfordiana   *Brodiaea crocea*

  ssp. *hispidula*   *Calochortus coeruleus*

Arenaria howellii   var. *maweanus*

A. rosei   *C. elegans* var. *nanus*

Arnica cernua   *Calycadenia ciliosa*

Asclepias solanoana   *Calycanthus occidentalis*

Astragalus breweri   *Calyptridium quadripetalum*

A. rattanii var. *jepsonianus*   *C. umbellatum* (= *Spraguea*

A. whitneyi var. *siskiyouensis*   *umbellata*)

Brodiaea stellaris   *Campanula scabrella*

Calochortus coeruleus   *Carex mendocinensis*

  var. *fimbriatus*   *Chamaecyparis lawsoniana*

C. greenei   *Convolvulus malacophyllus*

C. vestae   *Cordylanthus pilosus* ssp.

Castilleja brevilobata   *pilosus* (= ssp. *diffusus* of

C. miniata ssp. *elata*   Munz, in part)

Ceanothus pumilus   *C. tenuis* ssp. *viscidus*

Chlorogalum pomeridianum   *Cryptantha excavata*

  var. *minus*   *C. subretusa*

Collinsia greenei   *Cupressus bakeri*

Collomia diversifolia   *C. bakeri* ssp. *matthewsii*

Dentaria gemmata   *Cypripedium californicum*

D. pachystigma var. *dissectifolia*   *Delphinium decorum* ssp. *tracyi*

Dicentra oregona   *Dentaria tenella* var. *palmata*

Epilobium rigidum   *Draba howellii*

Eriogonum alpinum   *Epilobium obcordatum* var. *laxum*

E. congdonii   *E. oreganum*

E. kelloggii   *Eriogonum pyrolaefolium*

E. *libertini*

E. *siskiyouense*

E. *ternatum*

Erythronium californicum

E. *citrinum*

E. *hendersonii*

Fritillaria glauca

F. *purdyi*

F. *recurva* var. *coccinea*

Galium ambiguum

G. *ambiguum* var. *siskiyouense*

Gentiana bisetacea

G. *setigera*

Haplopappus ophitidis

H. *racemosus* ssp. *congestus*

Hesperolinon adenophyllum

H. *didymocarpum*

H. *tehamense*

Hieracium bolanderi

Horkelia sericata

Juniperus communis var. *jackii*

Lewisia stebbinsii

Lilium bolanderi

L. *kelloggii*

Linanthus dichotomus
ssp. *meridianus*

Lomatium ciliolatum

L. *ciliolatum* var. *hooveri*

L. *engelmannii*

L. *howellii*

L. *marginatum* var. *purpureum*

L. *tracyi*

Lupinus lapidicola

Perideridia leptocarpa

P. *oregana*

Eriophyllum lanatum
var. *lanceolatum*

Festuca tracyi

Frasera umquaensis

Fritillaria pluriflora

Galium ambiguum

Ivesia pickeringii

Juniperus communis
var. *saxatilis*

Lewisia cotyledon
ssp. *cotyledon*

L. *cotyledon* ssp. *heckneri*

L. *cotyledon* ssp. *howellii*

L. *oppositifolia*

Orthocarpus copelandii

Panicum thermale

Perideridia oregana

Phacelia leonis

P. *pringlei*

Picea breweriana

Pinus balfouriana
ssp. *balfouriana*

P. *jeffreyi*

Plagiobothrys lithocaryus

Polemonium chartaceum

Polystichum imbricans

P. *kruckebergii*

P. *scopulinum*

Pyrola picta ssp. *dentata*

Quercus vaccinifolia

Raillardella scabrida

Silene californica

S. *campanulata* ssp. *greenei*

S. *grayi*

S. *hookeri* var. *hookeri*

Phacelia corymbosa

P. dalesiana

P. greenei

Poa piperi

Polystichum lemmonii

Raillardella pringlei

R. scabrida

Rhamnus californica
  ssp. *crassifolia*

Rudbeckia californica
  var. *glauca*

Salix delnortensis

Sanicula peckiana

S. tracyi

Sedum laxum ssp. *eastwoodiae*

Senecio greenei

S. ligulifolius

Silene campanulata
  ssp. *campanulata*

S. campanulata ssp. *glandulosa*

S. hookeri ssp. *bolanderi*

Stipa lemmonii .ssp. *pubescens*

Streptanthus barbatus

S. drepanoides

S. howellii

Tauschia glauca

T. howellii

Thlaspi montanum
  var. *californicum*

Vancouveria chrysantha

Veronica copelandii

Viola cuneata

V. lobata ssp. *psychodes*

Tauschia glauca

Thlaspi montanum var. *montanum*

V. Coast Ranges (occur in portions of provinces I through IV)

Serpentine endemics   Serpentine indicators

Allium falcifolium   *Agropyron trachycaulum*

A. serratum   *Allium serratum*

Aquilegia eximia   *Angelica tomentosa*

Aspidotis carlotta-halliae   *Antirrhinum vexillo-calyculatum*

Calochortus umbellatus   *Aquilegia eximia*

Carex mendocinensis   *Calocedrus decurrens*

Ceanothus jepsonii   *Calochortus umbellatus*

Cirsium breweri   *Camissonia lacustris*

Convolvulus malacophyllus   *Carex serratodens*
   ssp. *collinus*   *Castilleja foliolosa*

Cupressus sargentii   *Chaenactis glabriuscula*

Hesperolinon disjunctum   var. *gracilenta*

Lessingia ramulosa   *Cirsium breweri*

Montia gypsophiloides   *Delphinium californicum*

Navarretia mitracarpa   ssp. *interius*

Parvisedum pentandrum   *D. hesperium*

Phacelia corymbosa   *Dodecatheon clevelandii*

Salix breweri   ssp. *patulum*

Streptanthus barbiger   *Eriogonum umbellatum*

S. breweri   ssp. *bahiaeforme*

Stylocline amphibola   *Festuca tracyi*

Zygadenus fontanus   *Fritillaria agrestis*

                                  F. biflora

                                  Garrya elliptica

                                  Hesperolinon breweri

                                  H. californicum

                                  H. clevelandii

                                  H. micranthum

                                  Lessingia ramulosa

                                  Lewisia rediviva

                                  L. triphylla

                                  Linum perenne ssp. *lewisii*

                                  Montia gypsophiloides

Navarretia pubescens
Phacelia californica
P. divaricata
P. egena
P. imbricata
Pinus attenuata
Plagiobothrys infectivus
Psoralea californica
Salix coulteri
Salvia sonomensis

Senecio eurycephalus
Streptanthus glandulosus
  ssp. *glandulosus*
Thelypodium flavescens
Umbellularia californica
Vancouveria planipetala
Viola ocellata
Zygadenus fontanus (= Z.
  micranthus **var.** *fontanus)*

VI. Sierra Nevada

Serpentine endemics

Allium sanbornii
Arabis constancei
A. suffrutescens **var.** *perstylosa*
Chlorogalum grandiflorum
Cryptantha mariposae
Githopsis pulchella
Lomatium congdonii
Lupinus spectabilis
Sedum albomarginatum
Senecio clevelandii
  var. *heterophyllus*
S. lewisrosei
Streptanthus polygaloides
S. tortuosus **var.** *optatus*

Serpentine indicators

Calycadenia mollis
C. oppositifolia
C. truncata **var.** *scabrella*
Calystegia stebbinsii
Clarkia arcuata
Cordylanthus tenuis
  ssp. *tenuis*
Dudleya cymosa **ssp.** *gigantea*
Eriophyllum confertiflorum
  var. *tanacetifolium*
E. nubigenum **var.** *congdonii*
Fremontodendron napensis
Fritillaria micrantha
Mimulus nasutus (M. pardalis
  in part)
Orthocarpus lacerus
Parvisedum pumilum

VII. Sierra Nevada and Coast Ranges

Serpentine endemics

Balsamorhiza macrolepis

Chlorogalum angustifolium

Eriogonum tripodum

Garrya congdonii

Helianthus bolanderi ssp. *exilis*

Lagophylla minor

Lithocarpus densiflora
   var. *echinoides*

Lomatium marginatum
   var. *marginatum*

Mondarella villosa ssp. *sheltonii*

Polygonum spergulariaeforme

Quercus durata

Sidalcea diploscypha

Trichostemma rubrisepalum

Serpentine indicators

Allium cratericola

Antirrhinum breweri

A. cornutum

Arabis suffrutescens

Arctostaphylos viscida

Arenaria douglasii

Aspidotis densa

Berberis pumila

Calochortus invenustus

C. nudus

Camissonia lacustris

Castilleja pruinosa

C. stenantha

Chaenactis glabriuscula

Chlorogalum angustifolium

Collinsia sparsiflora

Coreopsis stillmanii

Cryptantha milobakeri

Cupressus macnabiana

Darlingtonia californica

Delphinium patens

D. sonnei

Dicentra pauciflora

Eriogonum tripodum

Eriophyllum lanatum
   var. *achilleoides*

Fritillaria recurva
   var. *recurva*

Gentiana newberryi

Ivesia gordonii

Lewisia leana

Mimulus douglasii

M. nasutus

Muilla maritima

Odontostomum hartwegii

Phacelia egena

P. imbricata

Plagiobothrys greenei

P. shastensis

Poa tenerrima

Salvia sonomensis

Sedum obtusatum ssp. *obtusatum*

Senecio eurycephalus

Sidalcea diploscypha

Xerophyllum tenax

---

[1]Authorities for taxa given in appendices C and D.

LITERATURE CITED

ABRAMS, L.
   1923–1960.   Illustrated Flora of the Pacific States. 4 vol. Stanford, Calif.: Stanford University Press.

AMIDEI, G.
   1841.   Specie di piante osservate nei terreni serpentinosa. Atti Terza Riunione Scienz. It. (Florence), pp. 523-524.

ANTONOVICS, J., A.D. BRADSHAW, and R.G. TURNER
   1971.   Heavy metal tolerance in plants. Advances in Ecological Research 7:1-85.

AUBERY, L.E.
   1908.   Quicksilver resources of California. Calif. State Mining Bureau Bull. no. 27.

AYENSU, E.S., and R.A. DeFILIPPS
   1978.   Endangered and Threatened Plants of the United States. Washington: Smithsonian Institution and World Wildlife Fund, Inc.

BAILEY, D.K.
   1970.   Phytogeography and taxonomy of *Pinus* subsection Balfourianae. Ann. Missouri Bot. Gard. 57:210-249.

BAKER, H.G.
   1965.   Characteristics and modes of origin of weeds. *In* H.G. Baker and G.L. Stebbins, Jr. (eds.), The Genetics of Colonizing Species, pp. 147-172. New York: Academic Press.

BARBOUR, M.G. and J. MAJOR
   1977.   Terrestrial Vegetation of California. New York: Wiley.

BARNES, I., V.C. LaMARCHE, Jr., and G. HIMMELBERG
   1966.   Geochemical evidence of present-day serpentinization. Science 156:830-832.

BILLINGS, W.D.
   1950.   Vegetation and plant growth as affected by chemically altered rocks in the western Great Basin. Ecology 31:62-74.
   1952.   The environmental complex in relation to plant growth and distribution. Quarterly Rev. Biol. 27:251-265.

BOWERMAN, M.L.
   1944.   The Flowering Plants and Ferns of Mount Diablo. Berkeley: Gillick Press.

BRADLEY, W.W.
   1918.   Quicksilver resources of California. Calif. State Mining Bureau Bull. no. 78; 389 pp.

BREWER, W.H.
   1949.   Up and Down California in 1860–1864. Berkeley: University of California Press.

BROOKS, R.R.
   1972.   Geobotany and Biogeochemistry in Mineral Exploration. New York: Harper and Row.

CAIN, S.A.
    1944.   Foundations of Plant Geography. New York: Harper.
CALIFORNIA DIV. OF MINES AND GEOLOGY
    1958–1967.   Geologic Map of California (27 sheets). San Francisco.
CANNON, H.
    1960.   Botanical prospecting for ore deposits. Science 132:591-598.
CARLISLE, D. and G.B. CLEVELAND
    1958.   Plants as a guide to mineralization. Calif. Div. of Mines and Geology Special Report no.
             50.
CHALLINOR, J.
    1967.   A Dictionary of Geology, 3rd ed. New York: Oxford University Press.
CLAUSEN, J., D.D. KECK, and W.M. HEISEY
    1940.   Experimental studies on the nature of species. I. Effect of varied environments on western
             North American plants. Washington: Carnegie Inst. publ. no. 520.
    1948.   Experimental studies on the nature of species. III. Environmental responses of climatic
             races of *Achillea*. Washington: Carnegie Inst. publ. no. 581.
CODY, W.J.
    1983.   *Adiantum pedatum* ssp. *calderi*, a new subspecies in northeastern North America. Rho-
             dora 85:93-96.
COLEMAN, R.G.
    1967.   Low-temperature reaction zones and alpine ultramafic rocks of California, Oregon and
             Washington. U.S. Geol. Survey Bull. 1247:1-49.
    1977.   Ophiolites. Ancient Oceanic Lithosphere? Berlin: Springer-Verlag.
COOPER, W.S.
    1922.   The broad-leaved sclerophyll vegetation of California. Washington: Carnegie Inst. publ.
             no. 319:1-124.
FAUST, G.T. and J.J. FAHEY
    1962.   The serpentine-group minerals. U.S. Geol. Survey Prof. Paper no. 384-A:1-92.
FERLATTE, W.J.
    1974.   A Flora of the Trinity Alps of Northern California. Berkeley and Los Angeles: University
             of California Press.
FORDE, M.B., and D.G. FARIS
    1962.   Effect of introgression on the serpentine endemism of *Quercus durata*. Evolution 16:338-
             347.
FOWELLS, H.A.
    1965.   Silvics of Forest Trees of the United States. Washington: USDA Agric. Handbook no.
             271.
GANKIN, R., and J. MAJOR
    1964.   *Arctostaphylos myrtifolia*, its biology and relationship to the problem of endemism. Ecol-
             ogy 45:792-808.
GORDON, A., and C.B. LIPMAN
    1926.   Why are serpentine and other magnesian soils infertile? Soil Sci. 22:291-302.
GOTTLIEB, L.D.
    1973.   Genetic differentiation, sympatric speciation and the origin of a diploid species of *Stepha-
             nomeria*. Amer. J. Bot. 60:545-553.
GRAY, A., W.H. BREWER, and S. WATSON
    1880.   Botany. vols. 1 and 2. *In* J.D. Whitney, Geological Survey of California. Boston: Little,
             Brown.
GRAY, J.T.
    1979.   The vegetation of two California mountain slopes. Madrono 25:177-185.

GREENE, E.L.
  1904. Certain west American Cruciferae. Leaflets of Botanical Observation and Criticism no. 1:81-90.
GRIFFIN, J.R.
  1965. Digger pine seedling response to serpentinite and non-serpentinite soil. Ecology 46:801-807.
  1974. A strange pine and cedar forest in San Benito County. Fremontia 2:11-15.
  1975. Plants of the highest Santa Lucia and Diablo Range peaks, California. USDA Forest Service Res. Paper no. PSW-110.
GRIFFIN, J.R., and W.B. CRITCHFIELD
  1972. Distribution of forest trees in California. USDA Forest Service Res. Paper no. PSW-82.
HANES, T.L.
  1977. California chaparral. *In* M.G. Barbour and J. Major (eds.), Terrestrial Vegetation of California. New York: Wiley.
HARDHAM, C.B.
  1962. The Santa Lucia *Cupressus sargentii* groves and their associated northern hydrophilous and endemic species. Madrono 16:173-178.
HARSHBERGER, J.W.
  1911 Phytogeographic Survey of North America. Leipzig: W. Engleman.
HOFFMAN, F.W.
  1952. Studies in *Streptanthus*: A new *Streptanthus* complex in California. Madrono 11:189-220.
HOOVER, R.F.
  1970. The vascular plants of San Luis Obispo County, California. Berkeley and Los Angeles: University of California Press.
HOWELL, J.T.
  1970. Marin Flora, 2nd ed. Berkeley and Los Angeles: University of California Press.
JAFFRE, T., R.R. BROOKS, J. LEE, and R.D. REEVES
  1976. *Sebertia acuminata*: A hyperaccumulator of nickel from New Caledonia. Science 193:579-580.
JAFFRE, T., R.R. BROOKS, and J.M. TROW
  1979 Hyperaccumulation of nickel by *Geissosis* species. Plant and Soil 51:157-162.
JENNINGS, C.W., and R.G. STRAND
  1960. Ukiah sheet, Geologic Map of California. San Francisco: Calif. Div. of Mines and Geology.
JENNY, H.
  1980. The Soil Resource: Origin and Behavior. Berlin and New York: Springer-Verlag.
JEPSON, W.L.
  1909, 1936, 1943. A Flora of California, vols. 1, 2, and 3. Berkeley: Associated Students Store, University of California.
  1925. A Manual of the Flowering Plants of California. Berkeley: Associated Students Store, University of California.
JOHNSON, M.P., A.D. KEITH, and P.R. EHRLICH
  1968. The population biology of the butterfly, *Euphydryas editha*. VII. Has *E. editha* evolved a serpentine race? Evolution 22:422-423.
JONES, M.B., W.A. WILLIAMS, and J.E. RUCKMAN
  1977. Fertilization of *Trifolium subterraneum* L. growing on serpentine soils. Soil Sci. Soc. Amer. J. 41:87-89.
KOENIG, J.B.
  1963. Santa Rosa sheet, Geologic Map of California. San Francisco: Calif. Div. of Mines and Geology.

KOENIGS, R.L., W.A. WILLIAMS, and M.B. JONES
   1982.   Factors affecting vegetation on a serpentine soil. I. Principal components analysis of vegetation data. Hilgardia 50:1-14.
KOENIGS, R.L., W.A. WILLIAMS, M.B. JONES, and A. WALLACE
   1982.   Factors affecting vegetation on a serpentine soil. II. Chemical composition of foliage and soil. Hilgardia 50:15-26.
KRAUSE, W.
   1958.   Andere Bodenspezialisten. Handb. Pflanzenphysiol. 4:755-806.
KRUCKEBERG, A.R.
   1951.   Intraspecific variability in the response of certain native plant species to serpentine soil. Amer. J. Bot. 38:408-419.
   1954.   The ecology of serpentine soils: A symposium. III. Plant species in relation to serpentine soils. Ecology 35:267-274.
   1957.   Variation in fertility of hybrids between isolated populations of the serpentine species, *Streptanthus glandulosus* Hook. Evolution 11:185-211.
   1958.   The taxonomy of the species complex, *Streptanthus glandulosus* Hook. Madrono 14:217-248.
   1967.   Ecotypic response to ultramafic soils by some plant species of northwestern United States. Brittonia 19:133-151.
   1969a.   Soil diversity and the distribution of plants, with examples from western North America. Madrono 20:129-154.
   1969b.   Plant life on serpentinite and other ferromagnesian rocks in northwestern North America. Syesis 2:15-114.
KUCHLER, A.W.
   1977.   Natural vegetation of California (map). *In* M. Barbour and J. Major (eds.), Terrestrial Vegetation of California. New York: Wiley.
LEISER, A.T.
   1957.   *Rhododendron occidentale* on alkaline soil. Rhododendron and Camellia Year Book, 1957, p. 47-51.
LEWIS, H.
   1962.   Catastrophic selection as a factor in speciation. Evolution 16:257-271.
LLOYD, R.M.
   1965.   A new species of *Fremontodendron* (Sterculiaceae) from the Sierra Nevada foothills, California. Brittonia 17:382-384.
LYON, G.L., P.J. PETERSON, R.R. BROOKS, and G.W. BUTLER
   1971.   Calcium, magnesium and trace elements in New Zealand serpentine flora. J. Ecol. 59:421-429.
MADHOK, O.P., and R.B. WALKER
   1969.   Magnesium nutrition of two species of sunflower. Plant Physiol. 44:1016-1022.
MAIN, J.L.
   1974.   Differential responses to magnesium and calcium by native populations of *Agropyron spicatum*. Amer. J. Bot. 61:931-937.
MARTIN, W.E., J. VLAMIS, and N.W. STICE
   1953.   Field correction of calcium deficiency on a serpentine soil. Agronomy J. 45:204-208.
MASON, H.L.
   1946a.   The edaphic factor in narrow endemism. I. The nature of environmental influences. Madrono 8:209-226.
   1946b.   The edaphic factor in narrow endemism. II. The geographic occurrence of plants of highly restricted patterns of distribution. Madrono 8:241-257.

McMILLAN, C.
    1956.   The edaphic restriction of *Cupressus* and *Pinus* in the Coast Ranges of central California. Ecol. Monogr. 26:177-212.
McNAUGHTON, S.J.
    1968.   Structure and function in California grasslands. Ecology 49:962-972.
McNAUGHTON, S.J., T.C. FOLSOM, T. LEE, F. PARK, C. PRIVE, D. ROEDER, J. SCHMITS, and C. STOCKWELL
    1974.   Heavy metal tolerance in *Typha latifolia* without the evolution of tolerant races. Ecology 55:1163-1165.
MORRISON, J.L.
    1941.   A monograph of the section *Euclisia* Nutt. of *Streptanthus* Nutt. Ph.D. dissertation, University of California, Berkeley.
MUNZ, P.A., and D.D. KECK
    1959.   A California Flora. Berkeley and Los Angeles: University of California Press.
NEILSON, J.A.
    1977.   A report to Shell Oil Company in regards to observations on populations of the *Streptanthus morrisonii* complex in the central and southern Mayacamas Mts., Lake, Sonoma, and Napa counties, California. Ecoview, Environmental Consultants (mimeo).
NORRIS, R.M., and R.W. WEBB
    1976.   Geology of California. New York. Wiley.
NOVAK, F.A.
    1928.   Quelques remarques relatives au probleme de la vegetation sur les terrains serpentiniques. Preslia 6:42-71.
ORNDUFF, R.
    1974.   An Introduction to California Plant Life. Berkeley and Los Angeles: University of California Press.
PANCIC, J.
    1859.   Die Flora der Serpentinberge in Mittel-Serbien. Verh. zool. botan. Gesellschaft Wien 9:139-150.
PARSONS, R.F.
    1968.   The significance of growth-rate comparisons for plant ecology. Amer. Nat. 102:595-597.
PETERSON, P.J.
    1971.   Unusual accumulations of elements by plants and animals. Science Progress (Oxford) 59:505-526.
PICHI-SERMOLLI, R.
    1948.   Flora e vegetazione delle serpentine e delle altre ofioliti dell'alta valle del Trevere (Toscana). Webbia 6:1-380.
PROCTOR, J.
    1970.   Magnesium as a toxic element. Nature 227:742-743.
PROCTOR, J., and K. WHITTEN
    1971.   A population of the valley pocket gopher on a serpentine soil. Amer. Midland Nat. 85:517-521.
PROCTOR, J., and S.R.J. WOODELL
    1975.   The ecology of serpentine soils. Adv. Ecol. Res. 9:255-365.
RAI, D., G.H. SIMONSON, and C.T. YOUNGBERG
    1970.   Serpentine-derived soils in watershed and forest management. Report to U.S. Dept. of Interior, Bureau of Land Management. Dept. of Soils, Oregon State Univ., Corvallis (mimeo).
RAVEN, P.H.
    1964.   Catastrophic selection and edaphic endemism. Evolution 18:336-338.

RAVEN, P.H. and D.I. AXELROD
1978.   Origin and relationships of the California flora. Univ. Calif. Publs. in Botany 72:1-134.
REEVES, R.D., R.R. BROOKS, AND R.M. MACFARLANE
1981.   Nickel uptake by Californian *Streptanthus* and *Caulanthus* with particular reference to the hyperaccumulator, *S. polygaloides* Gray (Brassicaceae). Amer. J. Bot. 68:708-712.
RITTER-STUDNICKA, H.
1968.   Die Serpentinomorphosen der Flora Bosniens. Botanische Jarb. 88:443-465.
ROBINSON, W.O., G. EDGINGTON, and H.G. BYERS
1935.   Chemical studies of infertile soils derived from rocks high in magnesium and generally high in chromium and nickel. USDA Tech. Bull. no. 471:1-28.
RODMAN, J.E., A.R. KRUCKEBERG, and I.A. AL-SHEBAZ
1981.   Chemotaxonomic diversity and complexity in seed glucosinolates of *Caulanthus* and *Streptanthus* (Cruciferae). Systematic Bot. 6:197-222.
RUNE, O.
1953.   Plant life on serpentines and related rocks in the north of Sweden. Acta phytogeographica Suecica 31:1-139.
SHAPIRO, A.M.
1981.   Egg-mimics of *Streptanthus* (Cruciferae) deter oviposition by *Pieris sisymbrii* (Lepidoptera: Pieridae). Oecologia 48:142-143.
SHARSMITH, H.
1945.   Flora of the Mount Hamilton Range of California. Amer. Midland Nat. 34:289-367.
1961.   The genus *Hesperolinon* (Linaceae). Univ. Calif. Publs. in Botany 32:235-314.
SMITH, J.P., Jr., R.J. COLE, and J.O. SAWYER, Jr., in collaboration with W.R. POWELL
1980.   Inventory of rare and endangered vascular plants of California. California Native Plant Society (Berkeley) Special Publ. no. 1, 2nd ed.; 115 pp.
STEBBINS, G.L., Jr.
1942.   The genetic approach to problems of rare and endemic species. Madrono 6:241-272.
STEBBINS, G.L.
1980.   Rarity of plant species: A synthetic viewpoint. Rhodora 82:77-86.
STEBBINS, G.L., Jr., and J. MAJOR
1965.   Endemism and speciation in the California flora. Ecol. Monogr. 35:1-35.
STEBBINS, R.
1949.   Speciation in salamanders of the plethodontid genus *Ensatina*. Univ. Calif. Publs. in Zool. 48:377-526.
STRAND, R.G.
1962.   Redding sheet, Geologic Map of California. San Francisco: Calif. Div. of Mines and Geology.
TADROS, T.M.
1957.   Evidence of the presence of an edapho-biotic factor in the problem of serpentine tolerance. Ecology 38:14-23.
UNGER, F.
1836.   Ueber den Einfluss des Bodens auf die Verteilung der Gewaechse, nachgewiesen in der Vegetation des nordostlichen Tirols. Vienna: Rohrmann und Schweigerd.
USDA SOIL SURVEY STAFF
1975.   Soil Taxonomy. USDA Agric. Handbook no. 436.
VLAMIS, J.
1949.   Growth of lettuce and barley as influenced by degree of Ca saturation of soil. Soil Sci. 67:453-466.

VLAMIS, J., and H. JENNY
  1948.   Calcium deficiency in serpentine soils as revealed by absorbent technique. Science
            107:549-551.
VOGL, R.J.
  1973.   Ecology of knobcone pine in the Santa Ana Mountains, California. Ecol. Monogr.
            43:125-143.
WALKER, R.B.
  1948.   Molybdenum deficiency in serpentine barren soils. Science 108:473-475.
  1954.   Factors affecting plant growth on serpentine soils. *In* R.H. Whittaker et al., The ecology
            of serpentine soils: A symposium. Ecology 35:258-266.
WALKER, R.B., H.M. WALKER, and P.R. ASHWORTH.
  1955.   Calcium–magnesium nutrition, with special reference to serpentine soils. Plant Physiol-
            ogy 30:214-221.
WARING, R.H., and J. MAJOR
  1964.   Some vegetation of the California coastal redwood region in relation to gradients of mois-
            ture, nutrients, light and temperature. Ecol. Monogr. 34:167-215.
WELLS, P.V.
  1962.   Vegetation in relation to geological substratum and fire in the San Luis Obispo quadran-
            gle, California. Ecol. Monogr. 32:79-103.
WHITE, (' I)
  1971.   Vegetation–soil chemistry correlations in serpentine ecosystems. Ph.D. dissertation, Uni-
            versity of Oregon, Eugene.
WHITE, M.J.D.
  1978.   Modes of speciation. San Francisco: W.H. Freeman.
WHITNEY, J.D.
  1865.   Geology, vol. 1. Geological Survey of California. Philadelphia: Caxton Press.
WHITTAKER, R.H.
  1954.   IV. The vegetational response to serpentine soils. *In* R.H. Whittaker et al., The ecology
            of serpentine soils: A symposium. Ecology 35:275-288.
  1960.   Vegetation of the Siskiyou Mountains, Oregon and California. Ecol. Monogr. 30:279-
            338.
  1975.   Communities and ecosystems, 2nd ed. New York: Macmillan.
WHITTAKER, R.H., R.B. WALKER, and A.R. KRUCKEBERG.
  1954.   The ecology of serpentine soils: A symposium, pts. I–IV. Ecology 35:258-288.
WILD, H.
  1970.   Geobotanical anomalies in Rhodesia. 3. The vegetation of nickel-bearing soils. Kirkia 7,
            suppl.:1-62.
  1975.   Termites and the serpentines of the Great Dyke of Rhodesia. Trans. Rhodesia Scientific
            Assoc. 57:1-11.
WILD, H., and A.D. BRADSHAW
  1977.   The evolutionary effects of metalliferous and other anomalous soils in south central Af-
            rica. Evolution 31:282-293.
WILDMAN, W.E., M.L. JACKSON, and L.D. WHITTIG
  1968a.   Serpentinite rock dissolution as a function of carbon dioxide pressure in aqueous solution.
            Amer. Mineralogist 53:1252-1263.
  1968b.   Iron-rich montmorillonite formation in soils derived from serpentinite. Soil Sci. Soc.
            Amer. Proc. 32:787-794.
WRIGHT, R.D., and H.A. MOONEY
  1965.   Substrate-oriented distribution of bristlecone pine in the White Mountains of California.
            Amer. Midland Nat. 73:257-284.

WYLLIE, P.J. (ed.)
    1967.   Ultramafic and Related Rocks. New York: Wiley.
ZOBEL, B.
    1952.   Jeffrey pine in the south Coast Ranges of California. Madrono 11:283-284.
ZOBEL, D.G., and G.M. HAWK
    1980.   The environment of *Chamaecyparis lawsoniana*. Amer. Midland Nat. 103:280-297.

PLATES

FIGS. 20 to 32. Habit photos of selected serpentine plants.

FIG. 20. *Allium falcifolium*, southern Oregon. (Photo by M.F. Denton.)

FIG. 21. *Allium hoffmanii*, Mt. Lassic, Trinity County. (Photo by T. Nelson.)

FIG. 22. *Asclepias solanoana*, Dubakella Mountain, Trinity County. (Photo by T. Nelson.)

FIG. 23. *Berberis pumila*, Del Norte County. (Photo by M.F. Denton.)

FIG. 24. *Calochortus tiburonensis*, Tiburon peninsula, Marin County. (Photo by S. McKelvey.)

FIG. 25. *Castilleja neglecta,* Tiburon peninsula, Marin County. (Photo by W. Follette.)

FIG. 26. *Cupressus sargentii* in dense stand, Cuesta summit, San Luis Obispo County. (Photo by author.)

FIG. 27. *Cypripedium californicum*, Trinity Alps, Siskiyou County. (Photo by R. York.)

FIG. 28. *Darlingtonia californica*, along Illinois River, Josephine County, Oregon. (Photo by M.F. Denton.)

FIG. 29. *Eriogonum alpinum*, Mt. Eddy, Siskiyou County. (Photo by M.F. Denton.)

FIG. 30. *Lupinus sellulus* var. *humboldtensis*, Mt. Lassic, Trinity County. (Photo by T. Nelson.)

FIG. 31. *Polystichum lemmonii*, Mt. Eddy, Siskiyou County. (Photo by M.F. Denton.)

FIG. 32. *Streptanthus polygaloides*, Coulterville-Bagby area, Mariposa County. (Photo by author.)